Influence of Waves

STEM Road Map for Elementary School

Grade
1

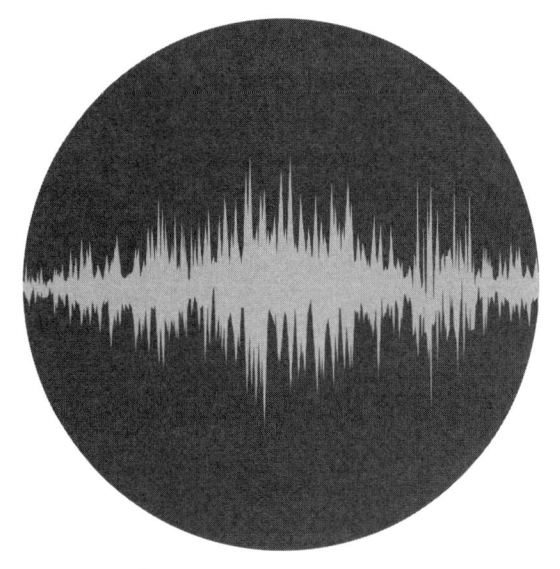

Influence of Waves

Grade
1

STEM Road Map
for Elementary School

Edited by Carla C. Johnson, Janet B. Walton, and
Erin Peters-Burton

press
National Science Teaching Association
Arlington, Virginia

National Science Teaching Association

Claire Reinburg, Director
Rachel Ledbetter, Managing Editor
Andrea Silen, Associate Editor
Jennifer Thompson, Associate Editor
Donna Yudkin, Book Acquisitions Manager

ART AND DESIGN
Will Thomas Jr., Director, cover and
    interior design
Himabindu Bichali, Graphic Designer, interior
    design

PRINTING AND PRODUCTION
Catherine Lorrain, Director

NATIONAL SCIENCE TEACHING ASSOCIATION
David L. Evans, Executive Director

1840 Wilson Blvd., Arlington, VA 22201
www.nsta.org/store
For customer service inquiries, please call 800-277-5300.

NSTA is committed to publishing material that promotes the best in inquiry-based science education. However, conditions of actual use may vary, and the safety procedures and practices described in this book are intended to serve only as a guide. Additional precautionary measures may be required. NSTA and the authors do not warrant or represent that the procedures and practices in this book meet any safety code or standard of federal, state, or local regulations. NSTA and the authors disclaim any liability for personal injury or damage to property arising out of or relating to the use of this book, including any of the recommendations, instructions, or materials contained therein.

PERMISSIONS
Book purchasers may photocopy, print, or e-mail up to five copies of an NSTA book chapter for personal use only; this does not include display or promotional use. Elementary, middle, and high school teachers may reproduce forms, sample documents, and single NSTA book chapters needed for classroom or noncommercial, professional-development use only. E-book buyers may download files to multiple personal devices but are prohibited from posting the files to third-party servers or websites, or from passing files to non-buyers. For additional permission to photocopy or use material electronically from this NSTA Press book, please contact the Copyright Clearance Center (CCC) (*www.copyright.com*; 978-750-8400). Please access *www.nsta.org/permissions* for further information about NSTA's rights and permissions policies.

Cataloging-in-Publication data for this book and the e-book are available from the Library of Congress.
ISBN: 978-1-68140-504-9
e-ISBN: 978-1-68140-505-6

The *Next Generation Science Standards* ("NGSS") were developed by twenty-six states, in collaboration with the National Research Council, the National Science Teaching Association and the American Association for the Advancement of Science in a process managed by Achieve, Inc. For more information go to *www.nextgenscience.org*.

CONTENTS

CONTENTS

ABOUT THE EDITORS AND AUTHORS

Dr. Carla C. Johnson is executive director of the William and Ida Friday Institute for Educational Innovation, associate dean, and professor of science education in the College of Education at North Carolina State University in Raleigh. She was most recently an associate dean, provost fellow, and professor of science education at Purdue University in West Lafayette, Indiana. Dr. Johnson serves as the director of research and evaluation for the Department of Defense–funded Army Educational Outreach Program (AEOP), a global portfolio of STEM education programs, competitions, and apprenticeships. She has been a leader in STEM education for the past decade, serving as the director of STEM Centers, editor of the *School Science and Mathematics* journal, and lead researcher for the evaluation of Tennessee's Race to the Top–funded STEM portfolio. Dr. Johnson has published over 100 articles, books, book chapters, and curriculum books focused on STEM education. She is a former science and social studies teacher and was the recipient of the 2013 Outstanding Science Teacher Educator of the Year award from the Association for Science Teacher Education (ASTE), the 2012 Award for Excellence in Integrating Science and Mathematics from the School Science and Mathematics Association (SSMA), the 2014 award for best paper on Implications of Research for Educational Practice from ASTE, and the 2006 Outstanding Early Career Scholar Award from SSMA. Her research focuses on STEM education policy implementation, effective science teaching, and integrated STEM approaches.

Dr. Janet B. Walton is a senior research scholar and the assistant director of evaluation for AEOP at North Carolina State University's William and Ida Friday Institute for Educational Innovation. She merges her economic development and education backgrounds to develop K–12 curricular materials that integrate real-life issues with sound cross-curricular content. Her research focuses on mixed methods research methodologies and collaboration between schools and community stakeholders for STEM education and problem- and project-based learning pedagogies. With this research agenda, she works to bring contextual STEM experiences into the classroom and provide students and educators with innovative resources and curricular materials.

Dr. Erin Peters-Burton is the Donna R. and David E. Sterling endowed professor in science education at George Mason University in Fairfax, Virginia. She uses her experiences from 15 years as an engineer and secondary science, engineering, and mathematics teacher to develop research projects that directly inform classroom practice in science and engineering. Her research agenda is based on the idea that all students should build self-awareness of how they learn science and engineering. She works to help students see themselves as "science-minded" and help teachers create classrooms that support student skills to develop scientific knowledge. To accomplish this, she pursues research projects that investigate ways that students and teachers can use self-regulated learning theory in science and engineering, as well as how inclusive STEM schools can help students succeed. During her tenure as a secondary teacher, she had a National Board Certification in Early Adolescent Science and was an Albert Einstein Distinguished Educator Fellow for NASA. As a researcher, Dr. Peters-Burton has published over 100 articles, books, book chapters, and curriculum books focused on STEM education and educational psychology. She received the Outstanding Science Teacher Educator of the Year award from ASTE in 2016 and a Teacher of Distinction Award and a Scholarly Achievement Award from George Mason University in 2012, and in 2010 she was named University Science Educator of the Year by the Virginia Association of Science Teachers.

Dr. Andrea R. Milner is the vice president and dean of academic affairs and an associate professor in the Teacher Education Department at Adrian College in Adrian, Michigan. A former early childhood and elementary teacher, Dr. Milner researches the effects constructivist classroom contextual factors have on student motivation and learning strategy use.

Dr. Tamara J. Moore is an associate professor of engineering education in the College of Engineering at Purdue University. Dr. Moore's research focuses on defining STEM integration through the use of engineering as the connection and investigating its power for student learning.

Dr. Vanessa B. Morrison is an associate professor in the Teacher Education Department at Adrian College. She is a former early childhood teacher and reading and language arts specialist whose research is focused on learning and teaching within a transdisciplinary framework.

Dr. Toni A. Sondergeld is an associate professor of assessment, research, and statistics in the School of Education at Drexel University in Philadelphia. Dr. Sondergeld's research concentrates on assessment and evaluation in education, with a focus on K–12 STEM.

ACKNOWLEDGMENTS

This module was developed as a part of the STEM Road Map project (Carla C. Johnson, principal investigator). The Purdue University College of Education, General Motors, and other sources provided funding for this project.

See *www.routledge.com/products/9781138804234* for more information about *STEM Road Map: A Framework for Integrated STEM Education.*

THE STEM ROAD MAP

BACKGROUND, THEORY, AND PRACTICE

OVERVIEW OF THE *STEM ROAD MAP CURRICULUM SERIES*

Carla C. Johnson, Erin Peters-Burton, and Tamara J. Moore

The *STEM Road Map Curriculum Series* was conceptualized and developed by a team of STEM educators from across the United States in response to a growing need to infuse real-world learning contexts, delivered through authentic problem-solving pedagogy, into K–12 classrooms. The curriculum series is grounded in integrated STEM, which focuses on the integration of the STEM disciplines—science, technology, engineering, and mathematics—delivered across content areas, incorporating the Framework for 21st Century Learning along with grade-level-appropriate academic standards.

The curriculum series begins in kindergarten, with a five-week instructional sequence that introduces students to the STEM themes and gives them grade-level-appropriate topics and real-world challenges or problems to solve. The series uses project-based and problem-based learning, presenting students with the problem or challenge during the first lesson, and then teaching them science, social studies, English language arts, mathematics, and other content, as they apply what they learn to the challenge or problem at hand.

Authentic assessment and differentiation are embedded throughout the modules. Each *STEM Road Map Curriculum Series* module has a lead discipline, which may be science, social studies, English language arts, or mathematics. All disciplines are integrated into each module, along with ties to engineering. Another key component is the use of STEM Research Notebooks to allow students to track their own learning progress. The modules are designed with a scaffolded approach, with increasingly complex concepts and skills introduced as students progress through grade levels.

The developers of this work view the curriculum as a resource that is intended to be used either as a whole or in part to meet the needs of districts, schools, and teachers who are implementing an integrated STEM approach. A variety of implementation formats are possible, from using one stand-alone module at a given grade level to using all five modules to provide 25 weeks of instruction. Also, within each grade band (K–2, 3–5, 6–8, 9–12), the modules can be sequenced in various ways to suit specific needs.

STANDARDS-BASED APPROACH

The *STEM Road Map Curriculum Series* is anchored in the *Next Generation Science Standards (NGSS)*, the *Common Core State Standards for Mathematics (CCSS Mathematics)*, the *Common Core State Standards for English Language Arts (CCSS ELA)*, and the Framework for 21st Century Learning. Each module includes a detailed curriculum map that incorporates the associated standards from the particular area correlated to lesson plans. The STEM Road Map has very clear and strong connections to these academic standards, and each of the grade-level topics was derived from the mapping of the standards to ensure alignment among topics, challenges or problems, and the required academic standards for students. Therefore, the curriculum series takes a standards-based approach and is designed to provide authentic contexts for application of required knowledge and skills.

THEMES IN THE *STEM ROAD MAP CURRICULUM SERIES*

The K–12 STEM Road Map is organized around five real-world STEM themes that were generated through an examination of the big ideas and challenges for society included in STEM standards and those that are persistent dilemmas for current and future generations:

- Cause and Effect

- Innovation and Progress

- The Represented World

- Sustainable Systems

- Optimizing the Human Experience

These themes are designed as springboards for launching students into an exploration of real-world learning situated within big ideas. Most important, the five STEM Road Map themes serve as a framework for scaffolding STEM learning across the K–12 continuum.

The themes are distributed across the STEM disciplines so that they represent the big ideas in science (Cause and Effect; Sustainable Systems), technology (Innovation and Progress; Optimizing the Human Experience), engineering (Innovation and Progress; Sustainable Systems; Optimizing the Human Experience), and mathematics (The Represented World), as well as concepts and challenges in social studies and 21st century skills that are also excellent contexts for learning in English language arts. The process of developing themes began with the clustering of the *NGSS* performance expectations and the National Academy of Engineering's grand challenges for engineering, which led to the development of the challenge in each module and connections of the module activities to the *CCSS Mathematics* and *CCSS ELA* standards. We performed these

mapping processes with large teams of experts and found that these five themes provided breadth, depth, and coherence to frame a high-quality STEM learning experience from kindergarten through 12th grade.

Cause and Effect

The concept of cause and effect is a powerful and pervasive notion in the STEM fields. It is the foundation of understanding how and why things happen as they do. Humans spend considerable effort and resources trying to understand the causes and effects of natural and designed phenomena to gain better control over events and the environment and to be prepared to react appropriately. Equipped with the knowledge of a specific cause-and-effect relationship, we can lead better lives or contribute to the community by altering the cause, leading to a different effect. For example, if a person recognizes that irresponsible energy consumption leads to global climate change, that person can act to remedy his or her contribution to the situation. Although cause and effect is a core idea in the STEM fields, it can actually be difficult to determine. Students should be capable of understanding not only when evidence points to cause and effect but also when evidence points to relationships but not direct causality. The major goal of education is to foster students to be empowered, analytic thinkers, capable of thinking through complex processes to make important decisions. Understanding causality, as well as when it cannot be determined, will help students become better consumers, global citizens, and community members.

Innovation and Progress

One of the most important factors in determining whether humans will have a positive future is innovation. Innovation is the driving force behind progress, which helps create possibilities that did not exist before. Innovation and progress are creative entities, but in the STEM fields, they are anchored by evidence and logic, and they use established concepts to move the STEM fields forward. In creating something new, students must consider what is already known in the STEM fields and apply this knowledge appropriately. When we innovate, we create value that was not there previously and create new conditions and possibilities for even more innovations. Students should consider how their innovations might affect progress and use their STEM thinking to change current human burdens to benefits. For example, if we develop more efficient cars that use by-products from another manufacturing industry, such as food processing, then we have used waste productively and reduced the need for the waste to be hauled away, an indirect benefit of the innovation.

The Represented World

When we communicate about the world we live in, how the world works, and how we can meet the needs of humans, sometimes we can use the actual phenomena to explain a concept. Sometimes, however, the concept is too big, too slow, too small, too fast, or too complex for us to explain using the actual phenomena, and we must use a representation or a model to help communicate the important features. We need representations and models such as graphs, tables, mathematical expressions, and diagrams because it makes our thinking visible. For example, when examining geologic time, we cannot actually observe the passage of such large chunks of time, so we create a timeline or a model that uses a proportional scale to visually illustrate how much time has passed for different eras. Another example may be something too complex for students at a particular grade level, such as explaining the p subshell orbitals of electrons to fifth graders. Instead, we use the Bohr model, which more closely represents the orbiting of planets and is accessible to fifth graders.

When we create models, they are helpful because they point out the most important features of a phenomenon. We also create representations of the world with mathematical functions, which help us change parameters to suit the situation. Creating representations of a phenomenon engages students because they are able to identify the important features of that phenomenon and communicate them directly. But because models are estimates of a phenomenon, they leave out some of the details, so it is important for students to evaluate their usefulness as well as their shortcomings.

Sustainable Systems

From an engineering perspective, the term *system* refers to the use of "concepts of component need, component interaction, systems interaction, and feedback. The interaction of subcomponents to produce a functional system is a common lens used by all engineering disciplines for understanding, analysis, and design." (Koehler, Bloom, and Binns 2013, p. 8). Systems can be either open (e.g., an ecosystem) or closed (e.g., a car battery). Ideally, a system should be sustainable, able to maintain equilibrium without much energy from outside the structure. Looking at a garden, we see flowers blooming, weeds sprouting, insects buzzing, and various forms of life living within its boundaries. This is an example of an ecosystem, a collection of living organisms that survive together, functioning as a system. The interaction of the organisms within the system and the influences of the environment (e.g., water, sunlight) can maintain the system for a period of time, thus demonstrating its ability to endure. Sustainability is a desirable feature of a system because it allows for existence of the entity in the long term.

In the STEM Road Map project, we identified different standards that we consider to be oriented toward systems that students should know and understand in the K–12 setting. These include ecosystems, the rock cycle, Earth processes (such as erosion, tectonics, ocean currents, weather phenomena), Earth-Sun-Moon cycles, heat transfer, and the interaction among the geosphere, biosphere, hydrosphere, and atmosphere. Students and teachers should understand that we live in a world of systems that are not independent of each other, but rather are intrinsically linked such that a disruption in one part of a system will have reverberating effects on other parts of the system.

Optimizing the Human Experience

Science, technology, engineering, and mathematics as disciplines have the capacity to continuously improve the ways humans live, interact, and find meaning in the world, thus working to optimize the human experience. This idea has two components: being more suited to our environment and being more fully human. For example, the progression of STEM ideas can help humans create solutions to complex problems, such as improving ways to access water sources, designing energy sources with minimal impact on our environment, developing new ways of communication and expression, and building efficient shelters. STEM ideas can also provide access to the secrets and wonders of nature. Learning in STEM requires students to think logically and systematically, which is a way of knowing the world that is markedly different from knowing the world as an artist. When students can employ various ways of knowing and understand when it is appropriate to use a different way of knowing or integrate ways of knowing, they are fully experiencing the best of what it is to be human. The problem-based learning scenarios provided in the STEM Road Map help students develop ways of thinking like STEM professionals as they ask questions and design solutions. They learn to optimize the human experience by innovating improvements in the designed world in which they live.

THE NEED FOR AN INTEGRATED STEM APPROACH

At a basic level, STEM stands for science, technology, engineering, and mathematics. Over the past decade, however, STEM has evolved to have a much broader scope and broader implications. Now, educators and policy makers refer to STEM as not only a concentrated area for investing in the future of the United States and other nations but also as a domain and mechanism for educational reform.

The good intentions of the recent decade-plus of focus on accountability and increased testing has resulted in significant decreases not only in instructional time for teaching science and social studies but also in the flexibility of teachers to promote authentic, problem solving–focused classroom environments. The shift has had a detrimental impact on student acquisition of vitally important skills, which many refer to as 21st century skills, and often the ability of students to "think." Further, schooling has become increasingly siloed into compartments of mathematics, science, English language arts, and social studies, lacking any of the connections that are overwhelmingly present in

the real world around children. Students have experienced school as content provided in boxes that must be memorized, devoid of any real-world context, and often have little understanding of why they are learning these things.

STEM-focused projects, curriculum, activities, and schools have emerged as a means to address these challenges. However, most of these efforts have continued to focus on the individual STEM disciplines (predominantly science and engineering) through more STEM classes and after-school programs in a "STEM enhanced" approach (Breiner et al. 2012). But in traditional and STEM enhanced approaches, there is little to no focus on other disciplines that are integral to the context of STEM in the real world. Integrated STEM education, on the other hand, infuses the learning of important STEM content and concepts with a much-needed emphasis on 21st century skills and a problem- and project-based pedagogy that more closely mirrors the real-world setting for society's challenges. It incorporates social studies, English language arts, and the arts as pivotal and necessary (Johnson 2013; Rennie, Venville, and Wallace 2012; Roehrig et al. 2012).

FRAMEWORK FOR STEM INTEGRATION IN THE CLASSROOM

The *STEM Road Map Curriculum Series* is grounded in the Framework for STEM Integration in the Classroom as conceptualized by Moore, Guzey, and Brown (2014) and Moore et al. (2014). The framework has six elements, described in the context of how they are used in the *STEM Road Map Curriculum Series* as follows:

1. The STEM Road Map contexts are meaningful to students and provide motivation to engage with the content. Together, these allow students to have different ways to enter into the challenge.

2. The STEM Road Map modules include engineering design that allows students to design technologies (i.e., products that are part of the designed world) for a compelling purpose.

3. The STEM Road Map modules provide students with the opportunities to learn from failure and redesign based on the lessons learned.

4. The STEM Road Map modules include standards-based disciplinary content as the learning objectives.

5. The STEM Road Map modules include student-centered pedagogies that allow students to grapple with the content, tie their ideas to the context, and learn to think for themselves as they deepen their conceptual knowledge.

6. The STEM Road Map modules emphasize 21st century skills and, in particular, highlight communication and teamwork.

All of the STEM Road Map modules incorporate these six elements; however, the level of emphasis on each of these elements varies based on the challenge or problem in each module.

THE NEED FOR THE *STEM ROAD MAP CURRICULUM SERIES*

As focus is increasing on integrated STEM, and additional schools and programs decide to move their curriculum and instruction in this direction, there is a need for high-quality, research-based curriculum designed with integrated STEM at the core. Several good resources are available to help teachers infuse engineering or more STEM enhanced approaches, but no curriculum exists that spans K–12 with an integrated STEM focus. The next chapter provides detailed information about the specific pedagogy, instructional strategies, and learning theory on which the *STEM Road Map Curriculum Series* is grounded.

REFERENCES

Breiner, J., M. Harkness, C. C. Johnson, and C. Koehler. 2012. What is STEM? A discussion about conceptions of STEM in education and partnerships. *School Science and Mathematics* 112 (1): 3–11.

Johnson, C. C. 2013. Conceptualizing integrated STEM education: Editorial. *School Science and Mathematics* 113 (8): 367–368.

Koehler, C. M., M. A. Bloom, and I. C. Binns. 2013. Lights, camera, action: Developing a methodology to document mainstream films' portrayal of nature of science and scientific inquiry. *Electronic Journal of Science Education* 17 (2).

Moore, T. J., S. S. Guzey, and A. Brown. 2014. Greenhouse design to increase habitable land: An engineering unit. *Science Scope* 37 (7): 51–57.

Moore, T. J., M. S. Stohlmann, H.-H. Wang, K. M. Tank, A. W. Glancy, and G. H. Roehrig. 2014. Implementation and integration of engineering in K–12 STEM education. In *Engineering in pre-college settings: Synthesizing research, policy, and practices*, ed. S. Purzer, J. Strobel, and M. Cardella, 35–60. West Lafayette, IN: Purdue Press.

Rennie, L., G. Venville, and J. Wallace. 2012. *Integrating science, technology, engineering, and mathematics: Issues, reflections, and ways forward.* New York: Routledge.

Roehrig, G. H., T. J. Moore, H. H. Wang, and M. S. Park. 2012. Is adding the *E* enough? Investigating the impact of K–12 engineering standards on the implementation of STEM integration. *School Science and Mathematics* 112 (1): 31–44.

STRATEGIES USED IN THE *STEM ROAD MAP CURRICULUM SERIES*

Erin Peters-Burton, Carla C. Johnson, Toni A. Sondergeld, and Tamara J. Moore

The *STEM Road Map Curriculum Series* uses what has been identified through research as best-practice pedagogy, including embedded formative assessment strategies throughout each module. This chapter briefly describes the key strategies that are employed in the series.

PROJECT- AND PROBLEM-BASED LEARNING

Each module in the *STEM Road Map Curriculum Series* uses either project-based learning or problem-based learning to drive the instruction. Project-based learning begins with a driving question to guide student teams in addressing a contextualized local or community problem or issue. The outcome of project-based instruction is a product that is conceptualized, designed, and tested through a series of scaffolded learning experiences (Blumenfeld et al. 1991; Krajcik and Blumenfeld 2006). Problem-based learning is often grounded in a fictitious scenario, challenge, or problem (Barell 2006; Lambros 2004). On the first day of instruction within the unit, student teams are provided with the context of the problem. Teams work through a series of activities and use open-ended research to develop their potential solution to the problem or challenge, which need not be a tangible product (Johnson 2003).

ENGINEERING DESIGN PROCESS

The *STEM Road Map Curriculum Series* uses engineering design as a way to facilitate integrated STEM within the modules. The engineering design process (EDP) is depicted in Figure 2.1 (p. 10). It highlights two major aspects of engineering design—problem scoping and solution generation—and six specific components of working toward a design: define the problem, learn about the problem, plan a solution, try the solution, test the solution, decide whether the solution is good enough. It also shows that communication

Figure 2.1. Engineering Design Process

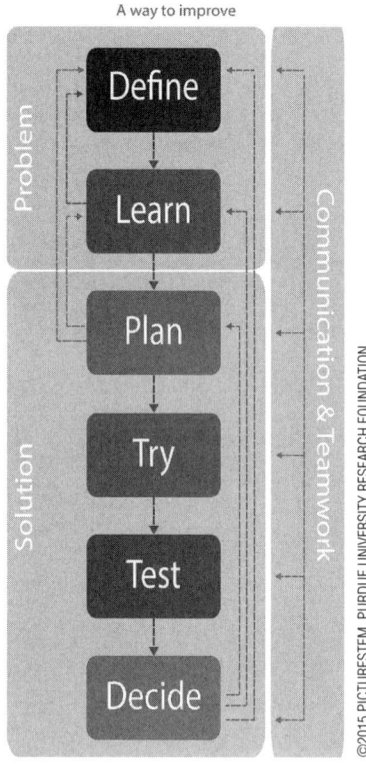

and teamwork are involved throughout the entire process. As the arrows in the figure indicate, the order in which the components of engineering design are addressed depends on what becomes needed as designers progress through the EDP. Designers must communicate and work in teams throughout the process. The EDP is iterative, meaning that components of the process can be repeated as needed until the design is good enough to present to the client as a potential solution to the problem.

Problem scoping is the process of gathering and analyzing information to deeply understand the engineering design problem. It includes defining the problem and learning about the problem. Defining the problem includes identifying the problem, the client, and the end user of the design. The client is the person (or people) who hired the designers to do the work, and the end user is the person (or people) who will use the final design. The designers must also identify the criteria and the constraints of the problem. The criteria are the things the client wants from the solution, and the constraints are the things that limit the possible solutions. The designers must spend significant time learning about the problem, which can include activities such as the following:

- Reading informational texts and researching about relevant concepts or contexts

- Identifying and learning about needed mathematical and scientific skills, knowledge, and tools

- Learning about things done previously to solve similar problems

- Experimenting with possible materials that could be used in the design

Problem scoping also allows designers to consider how to measure the success of the design in addressing specific criteria and staying within the constraints over multiple iterations of solution generation.

Solution generation includes planning a solution, trying the solution, testing the solution, and deciding whether the solution is good enough. Planning the solution includes generating many design ideas that both address the criteria and meet the constraints. Here the designers must consider what was learned about the problem during problem scoping. Design plans include clear communication of design ideas through media such as notebooks, blueprints, schematics, or storyboards. They also include details about the design, such as measurements, materials, colors, costs of materials, instructions for how things fit together, and sets of directions. Making the decision about which design idea to move forward involves considering the trade-offs of each design idea.

Once a clear design plan is in place, the designers must try the solution. Trying the solution includes developing a prototype (a testable model) based on the plan generated. The prototype might be something physical or a process to accomplish a goal. This component of design requires that the designers consider the risk involved in implementing the design. The prototype developed must be tested. Testing the solution includes conducting fair tests that verify whether the plan is a solution that is good enough to meet the client and end user needs and wants. Data need to be collected about the results of the tests of the prototype, and these data should be used to make evidence-based decisions regarding the design choices made in the plan. Here, the designers must again consider the criteria and constraints for the problem.

Using the data gathered from the testing, the designers must decide whether the solution is good enough to meet the client and end user needs and wants by assessment based on the criteria and constraints. Here, the designers must justify or reject design decisions based on the background research gathered while learning about the problem and on the evidence gathered during the testing of the solution. The designers must now decide whether to present the current solution to the client as a possibility or to do more iterations of design on the solution. If they decide that improvements need to be made to the solution, the designers must decide if there is more that needs to be understood about the problem, client, or end user; if another design idea should be tried; or if more planning needs to be conducted on the same design. One way or another, more work needs to be done.

Throughout the process of designing a solution to meet a client's needs and wants, designers work in teams and must communicate to each other, the client, and likely the end user. Teamwork is important in engineering design because multiple perspectives and differing skills and knowledge are valuable when working to solve problems. Communication is key to the success of the designed solution. Designers must communicate their ideas clearly using many different representations, such as text in an engineering notebook, diagrams, flowcharts, technical briefs, or memos to the client.

LEARNING CYCLE

The same format for the learning cycle is used in all grade levels throughout the STEM Road Map, so that students engage in a variety of activities to learn about phenomena in the modules thoroughly and have consistent experiences in the problem- and project-based learning modules. Expectations for learning by younger students are not as high as for older students, but the format of the progression of learning is the same. Students who have learned with curriculum from the STEM Road Map in early grades know

what to expect in later grades. The learning cycle consists of five parts—Introductory Activity/Engagement, Activity/Exploration, Explanation, Elaboration/Application of Knowledge, and Evaluation/Assessment—and is based on the empirically tested 5E model from BSCS (Bybee et al. 2006).

In the Introductory Activity/Engagement phase, teachers introduce the module challenge and use a unique approach designed to pique students' curiosity. This phase gets students to start thinking about what they already know about the topic and begin wondering about key ideas. The Introductory Activity/Engagement phase positions students to be confident about what they are about to learn, because they have prior knowledge, and clues them into what they don't yet know.

In the Activity/Exploration phase, the teacher sets up activities in which students experience a deeper look at the topics that were introduced earlier. Students engage in the activities and generate new questions or consider possibilities using preliminary investigations. Students work independently, in small groups, and in whole-group settings to conduct investigations, resulting in common experiences about the topic and skills involved in the real-world activities. Teachers can assess students' development of concepts and skills based on the common experiences during this phase.

During the Explanation phase, teachers direct students' attention to concepts they need to understand and skills they need to possess to accomplish the challenge. Students participate in activities to demonstrate their knowledge and skills to this point, and teachers can pinpoint gaps in student knowledge during this phase.

In the Elaboration/Application of Knowledge phase, teachers present students with activities that engage in higher-order thinking to create depth and breadth of student knowledge, while connecting ideas across topics within and across STEM. Students apply what they have learned thus far in the module to a new context or elaborate on what they have learned about the topic to a deeper level of detail.

In the last phase, Evaluation/Assessment, teachers give students summative feedback on their knowledge and skills as demonstrated through the challenge. This is not the only point of assessment (as discussed in the section on Embedded Formative Assessments), but it is an assessment of the culmination of the knowledge and skills for the module. Students demonstrate their cognitive growth at this point and reflect on how far they have come since the beginning of the module. The challenges are designed to be multidimensional in the ways students must collaborate and communicate their new knowledge.

STEM RESEARCH NOTEBOOK

One of the main components of the *STEM Road Map Curriculum Series* is the STEM Research Notebook, a place for students to capture their ideas, questions, observations, reflections, evidence of progress, and other items associated with their daily work. At the beginning of each module, the teacher walks students through the setup of the STEM

Research Notebook, which could be a three-ring binder, composition book, or spiral notebook. You may wish to have students create divided sections so that they can easily access work from various disciplines during the module. Electronic notebooks kept on student devices are also acceptable and encouraged. Students will develop their own table of contents and create chapters in the notebook for each module.

Each lesson in the *STEM Road Map Curriculum Series* includes one or more prompts that are designed for inclusion in the STEM Research Notebook and appear as questions or statements that the teacher assigns to students. These prompts require students to apply what they have learned across the lesson to solve the big problem or challenge for that module. Each lesson is designed to meaningfully refer students to the larger problem or challenge they have been assigned to solve with their teams. The STEM Research Notebook is designed to be a key formative assessment tool, as students' daily entries provide evidence of what they are learning. The notebook can be used as a mechanism for dialogue between the teacher and students, as well as for peer and self-evaluation.

The use of the STEM Research Notebook is designed to scaffold student notebooking skills across the grade bands in the *STEM Road Map Curriculum Series*. In the early grades, children learn how to organize their daily work in the notebook as a way to collect their products for future reference. In elementary school, students structure their notebooks to integrate background research along with their daily work and lesson prompts. In the upper grades (middle and high school), students expand their use of research and data gathering through team discussions to more closely mirror the work of STEM experts in the real world.

THE ROLE OF ASSESSMENT IN THE *STEM ROAD MAP CURRICULUM SERIES*

Starting in the middle years and continuing into secondary education, the word *assessment* typically brings grades to mind. These grades may take the form of a letter or a percentage, but they typically are used as a representation of a student's content mastery. If well thought out and implemented, however, classroom assessment can offer teachers, parents, and students valuable information about student learning and misconceptions that does not necessarily come in the form of a grade (Popham 2013).

The *STEM Road Map Curriculum Series* provides a set of assessments for each module. Teachers are encouraged to use assessment information for more than just assigning grades to students. Instead, assessments of activities requiring students to actively engage in their learning, such as student journaling in STEM Research Notebooks, collaborative presentations, and constructing graphic organizers, should be used to move student learning forward. Whereas other curriculum with assessments may include objective-type (multiple-choice or matching) tests, quizzes, or worksheets, we have intentionally avoided these forms of assessments to better align assessment strategies with teacher instruction and

student learning techniques. Since the focus of this book is on project- or problem-based STEM curriculum and instruction that focuses on higher-level thinking skills, appropriate and authentic performance assessments were developed to elicit the most reliable and valid indication of growth in student abilities (Brookhart and Nitko 2008).

Comprehensive Assessment System

Assessment throughout all STEM Road Map curriculum modules acts as a comprehensive system in which formative and summative assessments work together to provide teachers with high-quality information on student learning. Formative assessment occurs when the teacher finds out formally or informally what a student knows about a smaller, defined concept or skill and provides timely feedback to the student about his or her level of proficiency. Summative assessments occur when students have performed all activities in the module and are given a cumulative performance evaluation in which they demonstrate their growth in learning.

A comprehensive assessment system can be thought of as akin to a sporting event. Formative assessments are the practices: It is important to accomplish them consistently, they provide feedback to help students improve their learning, and making mistakes can be worthwhile if students are given an opportunity to learn from them. Summative assessments are the competitions: Students need to be prepared to perform at the best of their ability. Without multiple opportunities to practice skills along the way through formative assessments, students will not have the best chance of demonstrating growth in abilities through summative assessments (Black and Wiliam 1998).

Embedded Formative Assessments

Formative assessments in this module serve two main purposes: to provide feedback to students about their learning and to provide important information for the teacher to inform immediate instructional needs. Providing feedback to students is particularly important when conducting problem- or project-based learning because students take on much of the responsibility for learning, and teachers must facilitate student learning in an informed way. For example, if students are required to conduct research for the Activity/Exploration phase but are not familiar with what constitutes a reliable resource, they may develop misconceptions based on poor information. When a teacher monitors this learning through formative assessments and provides specific feedback related to the instructional goals, students are less likely to develop incomplete or incorrect conceptions in their independent investigations. By using formative assessment to detect problems in student learning and then acting on this information, teachers help move student learning forward through these teachable moments.

Formative assessments come in a variety of formats. They can be informal, such as asking students probing questions related to student knowledge or tasks or simply observing students engaged in an activity to gather information about student skills. Formative assessments can also be formal, such as a written quiz or a laboratory practical. Regardless of the type, three key steps must be completed when using formative assessments (Sondergeld, Bell, and Leusner 2010). First, the assessment is delivered to students so that teachers can collect data. Next, teachers analyze the data (student responses) to determine student strengths and areas that need additional support. Finally, teachers use the results from information collected to modify lessons and create learning environments that reinforce weak points in student learning. If student learning information is not used to modify instruction, the assessment cannot be considered formative in nature.

Formative assessments can be about content, science process skills, or even learning skills. When a formative assessment focuses on content, it assesses student knowledge about the disciplinary core ideas from the *Next Generation Science Standards* (*NGSS*) or content objectives from *Common Core State Standards for Mathematics* (*CCSS Mathematics*) or *Common Core State Standards for English Language Arts* (*CCSS ELA*). Content-focused formative assessments ask students questions about declarative knowledge regarding the concepts they have been learning. Process skills formative assessments examine the extent to which a student can perform science and engineering practices from the *NGSS* or process objectives from *CCSS Mathematics* or *CCSS ELA*, such as constructing an argument. Learning skills can also be assessed formatively by asking students to reflect on the ways they learn best during a module and identify ways they could have learned more.

Assessment Maps

Assessment maps or blueprints can be used to ensure alignment between classroom instruction and assessment. If what students are learning in the classroom is not the same as the content on which they are assessed, the resultant judgment made on student learning will be invalid (Brookhart and Nitko 2008). Therefore, the issue of instruction and assessment alignment is critical. The assessment map for this book (found in Chapter 3) indicates by lesson whether the assessment should be completed as a group or on an individual basis, identifies the assessment as formative or summative in nature, and aligns the assessment with its corresponding learning objectives.

Note that the module includes far more formative assessments than summative assessments. This is done intentionally to provide students with multiple opportunities to practice their learning of new skills before completing a summative assessment. Note also that formative assessments are used to collect information on only one or two learning objectives at a time so that potential relearning or instructional modifications can focus on smaller and more manageable chunks of information. Conversely, summative assessments in the module cover many more learning objectives, as they are traditionally used as final markers of student learning. This is not to say that information collected from summative assessments cannot or should not be used formatively. If teachers find that gaps in student

learning persist after a summative assessment is completed, it is important to revisit these existing misconceptions or areas of weakness before moving on (Black et al. 2003).

SELF-REGULATED LEARNING THEORY IN THE STEM ROAD MAP MODULES

Many learning theories are compatible with the STEM Road Map modules, such as constructivism, situated cognition, and meaningful learning. However, we feel that the self-regulated learning theory (SRL) aligns most appropriately (Zimmerman 2000). SRL requires students to understand that thinking needs to be motivated and managed (Ritchhart, Church, and Morrison 2011). The STEM Road Map modules are student centered and are designed to provide students with choices, concrete hands-on experiences, and opportunities to see and make connections, especially across subjects (Eliason and Jenkins 2012; NAEYC 2016). Additionally, SRL is compatible with the modules because it fosters a learning environment that supports students' motivation, enables students to become aware of their own learning strategies, and requires reflection on learning while experiencing the module (Peters and Kitsantas 2010).

The theory behind SRL (see Figure 2.2) explains the different processes that students engage in before, during, and after a learning task. Because SRL is a cyclical learning process, the accomplishment of one cycle develops strategies for the next learning cycle. This cyclic way of learning aligns with the various sections in the STEM Road Map lesson plans on Introductory Activity/ Engagement, Activity/Exploration, Explanation, Elaboration/Application of Knowledge, and Evaluation/Assessment. Since the students engaged in a module take on much of the responsibility for learning, this theory also provides guidance for teachers to keep students on the right track.

The remainder of this section explains how SRL theory is embedded within the five sections of each module and points out ways to

Figure 2.2. SRL Theory

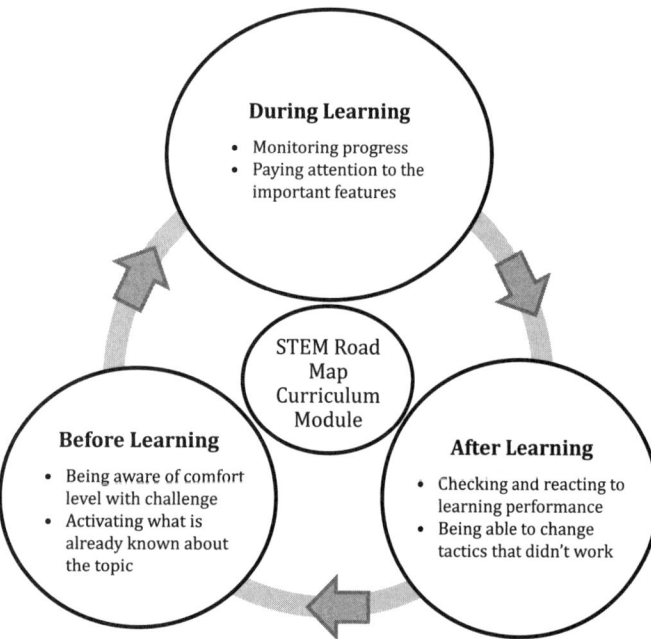

Source: Adapted from Zimmerman 2000.

support students in becoming independent learners of STEM while productively functioning in collaborative teams.

Before Learning: Setting the Stage

Before attempting a learning task such as the STEM Road Map modules, teachers should develop an understanding of their students' level of comfort with the process of accomplishing the learning and determine what they already know about the topic. When students are comfortable with attempting a learning task, they tend to take more risks in learning and as a result achieve deeper learning (Bandura 1986).

The STEM Road Map curriculum modules are designed to foster excitement from the very beginning. Each module has an Introductory Activity/Engagement section that introduces the overall topic from a unique and exciting perspective, engaging the students to learn more so that they can accomplish the challenge. The Introductory Activity also has a design component that helps teachers assess what students already know about the topic of the module. In addition to the deliberate designs in the lesson plans to support SRL, teachers can support a high level of student comfort with the learning challenge by finding out if students have ever accomplished the same kind of task and, if so, asking them to share what worked well for them.

During Learning: Staying the Course

Some students fear inquiry learning because they aren't sure what to do to be successful (Peters 2010). However, the STEM Road Map curriculum modules are embedded with tools to help students pay attention to knowledge and skills that are important for the learning task and to check student understanding along the way. One of the most important processes for learning is the ability for learners to monitor their own progress while performing a learning task (Peters 2012). The modules allow students to monitor their progress with tools such as the STEM Research Notebooks, in which they record what they know and can check whether they have acquired a complete set of knowledge and skills. The STEM Road Map modules support inquiry strategies that include previewing, questioning, predicting, clarifying, observing, discussing, and journaling (Morrison and Milner 2014). Through the use of technology throughout the modules, inquiry is supported by providing students access to resources and data while enabling them to process information, report the findings, collaborate, and develop 21st century skills.

It is important for teachers to encourage students to have an open mind about alternative solutions and procedures (Milner and Sondergeld 2015) when working through the STEM Road Map curriculum modules. Novice learners can have difficulty knowing what to pay attention to and tend to treat each possible avenue for information as equal (Benner 1984). Teachers are the mentors in a classroom and can point out ways for students to approach learning during the Activity/Exploration, Explanation, and

Elaboration/Application of Knowledge portions of the lesson plans to ensure that students pay attention to the important concepts and skills throughout the module. For example, if a student is to demonstrate conceptual awareness of motion when working on roller coaster research, but the student has misconceptions about motion, the teacher can step in and redirect student learning.

After Learning: Knowing What Works

The classroom is a busy place, and it may often seem that there is no time for self-reflection on learning. Although skipping this reflective process may save time in the short term, it reduces the ability to take into account things that worked well and things that didn't so that teaching the module may be improved next time. In the long run, SRL skills are critical for students to become independent learners who can adapt to new situations. By investing the time it takes to teach students SRL skills, teachers can save time later, because students will be able to apply methods and approaches for learning that they have found effective to new situations. In the Evaluation/Assessment portion of the STEM Road Map curriculum modules, as well as in the formative assessments throughout the modules, two processes in the after-learning phase are supported: evaluating one's own performance and accounting for ways to adapt tactics that didn't work well. Students have many opportunities to self-assess in formative assessments, both in groups and individually, using the rubrics provided in the modules.

The designs of the *NGSS* and *CCSS* allow for students to learn in diverse ways, and the STEM Road Map curriculum modules emphasize that students can use a variety of tactics to complete the learning process. For example, students can use STEM Research Notebooks to record what they have learned during the various research activities. Notebook entries might include putting objectives in students' own words, compiling their prior learning on the topic, documenting new learning, providing proof of what they learned, and reflecting on what they felt successful doing and what they felt they still needed to work on. Perhaps students didn't realize that they were supposed to connect what they already knew with what they learned. They could record this and would be prepared in the next learning task to begin connecting prior learning with new learning.

SAFETY IN STEM

Student safety is a primary consideration in all subjects but is an area of particular concern in science, where students may interact with unfamiliar tools and materials that may pose additional safety risks. It is important to implement safety practices within the context of STEM investigations, whether in a classroom laboratory or in the field. When you keep safety in mind as a teacher, you avoid many potential issues with the lesson while also protecting your students.

STEM safety practices encompass things considered in the typical science classroom. Ensure that students are familiar with basic safety considerations, such as wearing protective equipment (e.g., safety glasses or goggles and latex-free gloves) and taking care with sharp objects, and know emergency exit procedures. Teachers should learn beforehand the locations of the safety eyewash, fume hood, fire extinguishers, and emergency shut-off switch in the classroom and how to use them. Also be aware of any school or district safety policies that are in place and apply those that align with the work being conducted in the lesson. It is important to review all safety procedures annually.

STEM investigations should always be supervised. Each lesson in the modules includes teacher guidelines for applicable safety procedures that should be followed. Before each investigation, teachers should go over these safety procedures with the student teams. Some STEM focus areas such as engineering require that students can demonstrate how to properly use equipment in the maker space before the teacher allows them to proceed with the lesson.

The National Science Teaching Association (NSTA) provides a list of science rules and regulations, including standard operating procedures for lab safety, and a safety acknowledgment form for students and parents or guardians to sign. You can access these resources at *http://static.nsta.org/pdfs/SafetyInTheScienceClassroom.pdf*. In addition, NSTA's Safety in the Science Classroom web page (*www.nsta.org/safety*) has numerous links to safety resources, including papers written by the NSTA Safety Advisory Board.

Disclaimer: The safety precautions for each activity are based on use of the recommended materials and instructions, legal safety standards, and better professional practices. Using alternative materials or procedures for these activities may jeopardize the level of safety and therefore is at the user's own risk.

REFERENCES

Bandura, A. 1986. *Social foundations of thought and action: A social cognitive theory*. Englewood Cliffs, NJ: Prentice-Hall.

Barell, J. 2006. *Problem-based learning: An inquiry approach*. Thousand Oaks, CA: Corwin Press.

Benner, P. 1984. *From novice to expert: Excellence and power in clinical nursing practice*. Menlo Park, CA: Addison-Wesley.

Black, P., C. Harrison, C. Lee, B. Marshall, and D. Wiliam. 2003. *Assessment for learning: Putting it into practice*. Berkshire, UK: Open University Press.

Black, P., and D. Wiliam. 1998. Inside the black box: Raising standards through classroom assessment. *Phi Delta Kappan* 80 (2): 139–148.

Blumenfeld, P., E. Soloway, R. Marx, J. Krajcik, M. Guzdial, and A. Palincsar. 1991. Motivating project-based learning: Sustaining the doing, supporting learning. *Educational Psychologist* 26 (3): 369–398.

Brookhart, S. M., and A. J. Nitko. 2008. *Assessment and grading in classrooms*. Upper Saddle River, NJ: Pearson.

Bybee, R., J. Taylor, A. Gardner, P. Van Scotter, J. Carlson Powell, A. Westbrook, and N. Landes. 2006. *The BSCS 5E instructional model: Origins and effectiveness.* Colorado Springs, CO: BSCS.

Eliason, C. F., and L. T. Jenkins. 2012. *A practical guide to early childhood curriculum.* 9th ed. New York: Merrill.

Johnson, C. 2003. Bioterrorism is real-world science: Inquiry-based simulation mirrors real life. *Science Scope* 27 (3): 19–23.

Krajcik, J., and P. Blumenfeld. 2006. Project-based learning. In *The Cambridge handbook of the learning sciences,* ed. R. Keith Sawyer, 317–334. New York: Cambridge University Press.

Lambros, A. 2004. *Problem-based learning in middle and high school classrooms: A teacher's guide to implementation.* Thousand Oaks, CA: Corwin Press.

Milner, A. R., and T. Sondergeld. 2015. Gifted urban middle school students: The inquiry continuum and the nature of science. *National Journal of Urban Education and Practice* 8 (3): 442–461.

Morrison, V., and A. R. Milner. 2014. Literacy in support of science: A closer look at cross-curricular instructional practice. *Michigan Reading Journal* 46 (2): 42–56.

National Association for the Education of Young Children (NAEYC). 2016. Developmentally appropriate practice position statements. *www.naeyc.org/positionstatements/dap.*

Peters, E. E. 2010. Shifting to a student-centered science classroom: An exploration of teacher and student changes in perceptions and practices. *Journal of Science Teacher Education* 21 (3): 329–349.

Peters, E. E. 2012. Developing content knowledge in students through explicit teaching of the nature of science: Influences of goal setting and self-monitoring. *Science and Education* 21 (6): 881–898.

Peters, E. E., and A. Kitsantas. 2010. The effect of nature of science metacognitive prompts on science students' content and nature of science knowledge, metacognition, and self-regulatory efficacy. *School Science and Mathematics* 110: 382–396.

Popham, W. J. 2013. *Classroom assessment: What teachers need to know.* 7th ed. Upper Saddle River, NJ: Pearson.

Ritchhart, R., M. Church, and K. Morrison. 2011. *Making thinking visible: How to promote engagement, understanding, and independence for all learners.* San Francisco, CA: Jossey-Bass.

Sondergeld, T. A., C. A. Bell, and D. M. Leusner. 2010. Understanding how teachers engage in formative assessment. *Teaching and Learning* 24 (2): 72–86.

Zimmerman, B. J. 2000. Attaining self-regulation: A social-cognitive perspective. In *Handbook of self-regulation,* ed. M. Boekaerts, P. Pintrich, and M. Zeidner, 13–39. San Diego: Academic Press.

PART 2

INFLUENCE OF WAVES

STEM ROAD MAP MODULE

INFLUENCE OF WAVES
MODULE OVERVIEW

Andrea R. Milner, Vanessa B. Morrison, Janet B. Walton, Carla C. Johnson, and Erin Peters-Burton

THEME: Cause and Effect

LEAD DISCIPLINE: Science

MODULE SUMMARY

This module introduces students to the concept of waves as traveling disturbances that move through space and substances to transfer energy. Students apply their understanding and observations of waves in water to understanding sound waves to prepare them for an increasingly sophisticated understanding of the role of sound in communicating over distances. Students also investigate the properties of light and its role in communicating over distances. Students investigate how humans interact with sound and light. The focus on sound and light will lead to the final module challenge in which students use the steps of the engineering design process (EDP) to create musical instruments and work with lighting to stage a musical performance (adapted from Koehler, Bloom, and Milner 2015).

ESTABLISHED GOALS AND OBJECTIVES

At the conclusion of this module, students will be able to do the following:

- Know that there are different forms of waves

- Identify types of waves (e.g., waves in water and sound waves)

- Know that there are various sources of sound and light

- Describe how sound waves travel

- Explain that the body organs (e.g., eyes, ears, and skin) respond to sound and light

- Use technology to gather research information and communicate

- Describe and apply the EDP

- Design, construct, test, and evaluate models to demonstrate how humans experience and interact with sound and light

- Discuss concepts associated with sound and light when working on models to demonstrate how humans experience and interact with sound and light

- Identify impacts of sound and light on culture and society

CHALLENGE OR PROBLEM FOR STUDENTS TO SOLVE: SHOW ME THE WAVES CHALLENGE

Students will be challenged to design and create musical instruments for a musical show that will demonstrate how sound waves can be used to communicate over distances. Students will also incorporate light, another wave phenomenon, into their shows. Students will present musical performances that demonstrate how sound and light can be used to communicate and entertain.

CONTENT STANDARDS ADDRESSED IN THIS STEM ROAD MAP MODULE

A full listing with descriptions of the standards this module addresses can be found in Appendix C. Listings of the particular standards addressed within lessons are provided in a table for each lesson in Chapter 4.

STEM RESEARCH NOTEBOOK

Each student should maintain a STEM Research Notebook, which will serve as a place for students to organize their work throughout this module (see p. 12 for more general discussion on setup and use of the notebook). All written work in the module should be included in the notebook, including records of students' thoughts and ideas, fictional accounts based on the concepts in the module, and records of student progress through the EDP. The notebooks may be maintained across subject areas, giving students the opportunity to see that although their classes may be separated during the school day, the knowledge they gain is connected. The lesson plans for this module contain STEM Research Notebook Entry sections (numbered 1–33), and templates for each notebook entry are included in Appendix A.

Emphasize to students the importance of organizing all information in a Research Notebook. Explain to them that scientists and other researchers maintain detailed Research Notebooks in their work. These notebooks, which are crucial to researchers' work because they contain critical information and track the researchers' progress, are often considered legal documents for scientists who are pursuing patents or wish to provide proof of their discovery process.

MODULE LAUNCH

Following agreed-upon rules for discussions, launch the module by holding a class discussion about waves, asking students these questions:

- What are waves?

- Are there different types of waves?

- What kinds of waves are there?

- What makes waves?

- Where and when have you seen waves?

Students will share their ideas with the whole class. After students have shared their ideas about waves, show a video about ocean waves such as the drone video found at *www.wtkr.com/2014/01/17/this-drone-shot-video-of-surfers-in-hawaii-will-blow-your-mind*. Then, hold a class discussion about what students observed about waves in this video.

PREREQUISITE SKILLS FOR THE MODULE

Students enter this module with a wide range of preexisting skills, information, and knowledge. Table 3.1 provides an overview of prerequisite skills and knowledge that students are expected to apply in this module, along with examples of how they apply this knowledge throughout the module. Differentiation strategies are also provided for students who may need additional support in acquiring or applying this knowledge.

Table 3.1. Prerequisite Key Knowledge and Examples of Applications and Differentiation Strategies

Prerequisite Key Knowledge	Application of Knowledge by Students	Differentiation for Students Needing Additional Support
Science • Understanding cause and effect	*Science* • Determine in investigations how sound and light behave and interact with human organs.	*Science* • Provide demonstrations of cause and effect (e.g., dropping egg [cause] and observing breakage [effect]), emphasizing that cause is why something happens and effect is what happens. • Read aloud picture books to the class and have students identify cause-and-effect sequences.

Continued

Table 3.1. (*continued*)

Prerequisite Key Knowledge	Application of Knowledge by Students	Differentiation for Students Needing Additional Support
Mathematics • Number sense	*Mathematics* • Measure, calculate, compare, and evaluate numbers when investigating sound and light.	*Mathematics* • Model measurement techniques using standard and nonstandard units of measurement. • Read aloud nonfiction texts about temperature, rainfall, wind, and measurement. • Provide opportunities for students to practice measurement in a variety of settings (e.g., in the classroom and outdoors).
Language and Inquiry Skills • Visualizing • Making predictions • Recording ideas and information using words and pictures • Asking and responding to questions	*Language and Inquiry Skills* • Make and confirm or reject predictions. • Share thought processes through keeping a notebook, asking and responding to questions, and use of the engineering design process.	*Language and Inquiry Skills* • As a class, make predictions when reading fictional texts. • Model the process of using information and prior knowledge to use predictions. • Provide samples of notebook entries.
Speaking and Listening • Participating in group discussions	*Speaking and Listening* • Engage in collaborative group discussions in creating musical instruments and planning team Show Me the Waves presentations.	*Speaking and Listening* • Model speaking and listening skills. • Create a class list of good speaking and listening skills. • Create a class list of good collaboration practices. • Read picture books that feature collaboration and teamwork.

POTENTIAL STEM MISCONCEPTIONS

Students enter the classroom with a wide variety of prior knowledge and ideas, so it is important to be alert to misconceptions, or inappropriate understandings of foundational knowledge. These misconceptions can be classified as one of several types: "preconceived notions," opinions based on popular beliefs or understandings; "nonscientific beliefs," knowledge students have gained about science from sources outside the scientific community; "conceptual misunderstandings," incorrect conceptual models based on incomplete understanding of concepts; "vernacular misconceptions," misunderstandings of words based on their common use versus their scientific use; and "factual misconceptions," incorrect or imprecise knowledge learned in early life that remains unchallenged (NRC 1997, p. 28). Misconceptions must be addressed and dismantled for students to reconstruct their knowledge, and therefore teachers should be prepared to take the following steps:

- *Identify students' misconceptions.*

- *Provide a forum for students to confront their misconceptions.*

- *Help students reconstruct and internalize their knowledge, based on scientific models. (NRC 1997, p. 29)*

Keeley and Harrington (2010) recommend using diagnostic tools such as probes and formative assessment to identify and confront student misconceptions and begin the process of reconstructing student knowledge. Keeley's *Uncovering Student Ideas in Science* series contains probes targeted toward uncovering student misconceptions in a variety of areas and may be a useful resource for addressing student misconceptions in this module.

Some commonly held misconceptions specific to lesson content are provided with each lesson so that you can be alert for student misunderstanding of the science concepts presented and used during this module. The American Association for the Advancement of Science has also identified misconceptions that students frequently hold regarding various science concepts (see the links at *http://assessment.aaas.org/topics*).

SRL PROCESS COMPONENTS

Table 3.2 (p. 28) illustrates some of the activities in the Influence of Waves module and how they align with the self-regulated learning (SRL) process before, during, and after learning.

Table 3.2. SRL Process Components

Learning Process Components	Example From Influence of Waves Module	Lesson Number and Learning Component
	BEFORE LEARNING	
Motivates students	Students will experience waves in water by watching a video and sharing what they learned about waves as a whole class and in their STEM Research Notebooks.	Lesson 1, Introductory Activity/Engagement
Evokes prior learning	Students share what they know about waves from experience and respond to the following questions about waves: • What are waves? • Are there different types of waves? • What kinds of waves are there? • What causes waves? • Where and when have you seen waves?	Lesson 1, Introductory Activity/Engagement
	DURING LEARNING	
Focuses on important features	Students participate in the What's the Buzz? investigation by predicting, observing, and explaining how sound waves behave in a communication device.	Lesson 2, Activity/ Exploration
Helps students monitor their progress	After conducting an investigation about light's interaction with various materials, students consider how they could use their findings to use light in their musical performances. Teachers monitor students' responses and help students understand concepts if there are misconceptions.	Lesson 3, Explanation
	AFTER LEARNING	
Evaluates learning	Students demonstrate how humans experience and interact with sound waves and light waves through a Show Me the Waves musical performance. Students create guitars and use these guitars and the drums they created in a previous lesson, along with their voices and flashlights wrapped with colored materials, to present a musical performance that demonstrates how sound waves and light can be used to communicate and entertain.	Lesson 4, Activity/ Exploration
Takes account of what worked and what did not work	Students compare their predictions for the Singing Strings activity with actual results. Groups use what they learned from creating drums and guitars to create new instruments in the Incredible Instruments activity.	Lesson 4, Activity/ Exploration

STRATEGIES FOR DIFFERENTIATING INSTRUCTION WITHIN THIS MODULE

For the purposes of this curriculum module, differentiated instruction is conceptualized as a way to tailor instruction—including process, content, and product—to various student needs in your class. A number of differentiation strategies are integrated into lessons across the module. The problem- and project-based learning approach used in the lessons is designed to address students' multiple intelligences by providing a variety of entry points and methods to investigate the key concepts in the module (for example, investigating waves via scientific inquiry, literature, journaling, and collaborative design). Differentiation strategies for students needing support in prerequisite knowledge can be found in Table 3.1 (p. 25). You are encouraged to use information gained about student prior knowledge during introductory activities and discussions to inform your instructional differentiation. Strategies incorporated into this lesson include flexible grouping, varied environmental learning contexts, assessments, compacting, tiered assignments and scaffolding, and mentoring.

The following websites may be helpful resources for differentiated instruction:

- *http://steinhardt.nyu.edu/scmsAdmin/uploads/005/120/Culturally%20Responsive%20 Differentiated%20Instruction.pdf*

- *http://educationnorthwest.org/sites/default/files/12.99.pdf*

Flexible Grouping. Students work collaboratively in a variety of activities throughout this module. Grouping strategies you might employ include student-led grouping, grouping students according to ability level or common interests, grouping students randomly, or grouping them so that students in each group have complementary strengths (for instance, one student might be strong in mathematics, another in art, and another in writing).

Varied Environmental Learning Contexts. Students have the opportunity to learn in various contexts throughout the module, including alone, in groups, in quiet reading and research-oriented activities, and in active learning in inquiry and design activities. In addition, students learn in a variety of ways through doing inquiry activities, journaling, reading a variety of texts, watching videos, participating in class discussion, and conducting web-based research.

Assessments. Students are assessed in a variety of ways throughout the module, including individual and collaborative formative and summative assessments. Students have the opportunity to produce work via written text, oral and media presentations, and modeling.

Compacting. Based on student prior knowledge, you may wish to adjust instructional activities for students who exhibit prior mastery of a learning objective. Since student

work in science is largely collaborative throughout the module, this strategy may be most appropriate for mathematics, English language arts (ELA), or social studies activities. You may wish to compile a classroom database of research resources and supplementary readings for a variety of reading levels and on a variety of topics related to the module's topic to provide opportunities for students to undertake independent reading.

Tiered Assignments and Scaffolding. Based on your awareness of student ability, understanding of concepts, and mastery of skills, you may wish to provide students with variations on activities by adding complexity to assignments or providing more or fewer learning supports for activities throughout the module. For instance, some students may need additional support in identifying key search words and phrases for web-based research or may benefit from cloze sentence handouts to enhance vocabulary understanding. Other students may benefit from expanded reading selections and additional reflective writing or from working with manipulatives and other visual representations of mathematical concepts. You may also work with your school librarian to compile a classroom database of research resources and supplementary readings for different reading levels and on a variety of topics related to the module challenge to provide opportunities for students to undertake independent reading. You may find the following website on scaffolding strategies helpful: *www.edutopia.org/blog/scaffolding-lessons-six-strategies-rebecca-alber.*

Mentoring. As group design teamwork becomes increasingly complex throughout the module, you may wish to have a resource teacher, older student, or volunteer work with groups that struggle to stay on task and collaborate effectively.

STRATEGIES FOR ENGLISH LANGUAGE LEARNERS

Students who are developing proficiency in English language skills require additional supports to simultaneously learn academic content and the specialized language associated with specific content areas. WIDA (2012) has created a framework for providing support to these students and makes available rubrics and guidance on differentiating instructional materials for English language learners (ELLs). In particular, ELL students may benefit from additional sensory supports such as images, physical modeling, and graphic representations of module content, as well as interactive support through collaborative work. This module incorporates a variety of sensory supports and offers ongoing opportunities for ELL students to work with collaboratively. The focus on sound and light issues affords an opportunity for ELL students to share culturally diverse experiences with music and hearing and their experiences with visual impairments.

When differentiating instruction for ELL students, you should carefully consider the needs of these students as you introduce and use academic language in various language domains (listening, speaking, reading, and writing) throughout this module. To adequately differentiate instruction for ELL students, you should have an understanding

of the proficiency level of each student. The following five overarching preK–5 WIDA learning standards are relevant to this module:

- Standard 1: Social and Instructional Language. Focus on following directions, personal information, leisure activities, collaboration with peers.

- Standard 2: The Language of Language Arts. Focus on nonfiction, fiction, sequence of story, elements of story.

- Standard 3: The Language of Mathematics. Focus on basic operations, number sense, interpretation of data, standard and metric measurement tools.

- Standard 4: The Language of Science. Focus on light, scientific process, senses, sound.

- Standard 5: The Language of Social Studies. Focus on community workers, jobs and careers, representations of Earth (maps and globes).

SAFETY CONSIDERATIONS FOR THE ACTIVITIES IN THIS MODULE

The safety precautions associated with each investigation are based, in part, on the use of the recommended materials and instructions, legal safety standards, and better professional safety practices. Selection of alternative materials or procedures for these investigations may jeopardize the level of safety and therefore is at the user's own risk. Remember that an investigation includes three parts: (1) setup, in which you prepare the materials for students to use; (2) the actual hands-on investigation, in which students use the materials and equipment; and (3) cleanup, in which you or the students clean the materials and put them away for later use. The safety procedures for each investigation apply to all three parts. For more general safety guidelines, see the Safety in STEM section in Chapter 2 (p. 18).

We also recommend that you go over the safety rules that are included as part of the safety acknowledgment form with your students before beginning the first investigation. Once you have gone over these rules with your students, have them sign the safety acknowledgment form. You should also send the form home with students for parents or guardians to read and sign to acknowledge that they understand the safety procedures that must be followed by their children. A sample elementary safety acknowledgment form can be found on the National Science Teaching Association's Safety Portal at *http://static.nsta.org/pdfs/SafetyAcknowledgmentForm-ElementarySchool.pdf.*

DESIRED OUTCOMES AND MONITORING SUCCESS

The desired outcomes for this module are outlined in Table 3.3, along with suggested ways to gather evidence to monitor student success. For more specific details on desired outcomes, see the Established Goals and Objectives sections for the module (p. 23) and individual lessons.

Table 3.3. Desired Outcomes and Evidence of Success in Achieving Identified Outcomes

Desired Outcome	Evidence of Success	
	Performance Tasks	Other Measures
Students understand and can demonstrate their knowledge of how sound and light can be used to communicate and how humans interact with sound and light.	• Students complete a variety of investigations related to waves, sound, and light. • Student teams develop and present a musical performance that incorporates sound and light to demonstrate how sound and light can be used to communicate and how humans interact with sound and light. • Students each maintain a STEM Research Notebook that includes what they have learned, responses to questions, and observations.	Students are assessed using the Observation, STEM Research Notebook, and Participation Rubric.

ASSESSMENT PLAN OVERVIEW AND MAP

Table 3.4 provides an overview of the major group and individual *products* and *deliverables,* or things that students will produce in this module, that comprise the assessment for this module. See Table 3.5 for a full assessment map of formative and summative assessments in this module.

Table 3.4. Major Products and Deliverables in Lead Discipline for Groups and Individuals

Lesson	Major Group Products and Deliverables	Major Individual Products and Deliverables
1	• Waves in Motion investigation • Wave measurements	• STEM Research Notebook entries #1–8 • Lesson assessment
2	• What's the Buzz? communication devices • I Can Hear the Music instruments	• STEM Research Notebook entries #9–18 • Lesson assessment
3	• Lighting the Way investigation • Shady Shadow Puppets investigation • Shady Shadow Puppets show	• STEM Research Notebook entries #19–29 • Shadow image pictures • Partner description pictures • Lesson assessment
4	• Singing Strings activity • Incredible Instruments activity • Show Me the Waves musical performance	• STEM Research Notebook entries #30–33 • Lesson assessment

Table 3.5. Assessment Map for Influence of Waves Module

Lesson	Assessment	Group/Individual	Formative/Summative	Lesson Objective Assessed
1	Waves in Motion Investigation *performance task*	Group	Formative	• Demonstrate how waves in water move.
1	STEM Research Notebook *entries*	Individual	Formative	• Explain that waves transfer energy. • Demonstrate how waves in water move. • Identify properties of waves.
1	Wave drawing and descriptions *end of lesson assessment*	Individual	Summative	• Explain that waves transfer energy. • Demonstrate how waves in water move.

Continued

Table 3.5. (*continued*)

Lesson	Assessment	Group/ Individual	Formative/ Summative	Lesson Objective Assessed
2	STEM Research Notebook entries	Individual	Formative	• Understand that there are various sources of sound waves and identify sources of sound. • Demonstrate a basic understanding of the human anatomy associated with hearing. • Explain that sound can be used to communicate over distances. • Evaluate the influence sound waves have on culture and society.
2	What's the Buzz? *investigation*	Group	Formative	• Design, construct, test, and evaluate models that demonstrate how humans experience and interact with sound waves. • Explain how sound waves travel to reach human ears. • Explain that sound can be used to communicate over distances
2	I Can Hear the Music Instruments *investigation*	Group	Formative	• Design, construct, test, and evaluate models that demonstrate how humans experience and interact with sound waves. • Explain that sound can be used to communicate over distances.
2	Drawing and description of sources of sound and basic ear anatomy *end of lesson assessment*	Individual	Summative	• Understand that there are various sources of sound waves and identify sources of sound. • Demonstrate a basic understanding of the human anatomy associated with hearing.
3	STEM Research Notebook *entries*	Individual	Formative	• Identify the Sun as a natural source of light. • Describe how light reaches our eyes. • Understand how human eyes respond to light via a basic exploration of the human anatomy of sight. • Identify materials that are transparent, translucent, or opaque. • Identify several ways that humans experience and interact with light waves. • Explain how light can be used to communicate over distances.

Continued

Table 3.5. (*continued*)

Lesson	Assessment	Group/Individual	Formative/Summative	Lesson Objective Assessed
3	Lighting the Way *investigation*	Group	Formative	• Identify materials that are transparent, translucent, or opaque. • Explain concepts associated with light when working on models to demonstrate how humans experience and interact with light waves. • Explain how light can be used to communicate over distances.
3	Shady Shadow Puppets *investigation*	Group	Formative	• Identify several ways that humans experience and interact with light waves. • Describe and demonstrate how shadows are formed. • Explain how light can be used to communicate over distances.
3	Shady Shadow Puppets *performance task*	Group	Formative	• Describe and demonstrate how shadows are formed.
3	Shadow image pictures *performance task*	Individual	Formative	• Identify the Sun as a natural source of light. • Explain concepts associated with light when working on models to demonstrate how humans experience and interact with light waves. • Describe and demonstrate how shadows are formed.
3	Partner description pictures *performance task*	Individual	Formative	• Demonstrate an understanding of the use of descriptive words and imagery in art and literature.
3	Light drawings and descriptions *end of lesson assessment*	Individual	Summative	• Identify several ways that humans experience and interact with light waves. • Explain how light can be used to communicate over distances. • Identify materials that are transparent, translucent, or opaque.

Continued

Table 3.5. (*continued*)

Lesson	Assessment	Group/ Individual	Formative/ Summative	Lesson Objective Assessed
4	STEM Research Notebook *entries*	Individual	Formative	• Explain that the pitch of a sound is influenced by properties of the material through which the sound waves move. • Demonstrate their understanding of the engineering design process (EDP) by applying it to create musical instruments.
4	Singing Strings *performance task*	Group	Formative	• Explain that the pitch of a sound is influenced by properties of the material through which the sound waves move. • Explain that sound can be used to communicate over distance and create models that demonstrate this.
4	Incredible Instruments *performance task*	Group	Formative	• Demonstrate their understanding of the EDP by applying it to create musical instruments. • Explain that sound and light are used to communicate over distances and identify several examples of this.
4	Musical Show *performance task*	Group	Summative	• Explain that sound can be used to communicate over distances and create models that demonstrate this. • Explain that light can be used to communicate over distances and create a model that demonstrates this. • Demonstrate how humans can experience and interact with sound and light through a musical performance.
4	Sound and light drawing and descriptions *end of lesson assessment*	Individual	Summative	• Identify several ways that humans experience and interact with sound and light. • Explain that sound can be used to communicate over distances and create models that demonstrate this. • Explain that light can be used to communicate over distances and create a model that demonstrates this.

MODULE TIMELINE

Tables 3.6–3.10 (pp. 37–39) provide lesson timelines for each week of the module. The timelines are provided for general guidance only and are based on class times of approximately 30 minutes.

Table 3.6. STEM Road Map Module Schedule for Week One

Day 1	Day 2	Day 3	Day 4	Day 5
Lesson 1 *Let's Explore Waves!* • Launch the module with group discussion and video about waves in water. • Introduce STEM Research Notebooks. • Discuss and demonstrate energy.	*Lesson 1* *Let's Explore Waves!* • Conduct interactive read-aloud of *Waves: Energy on the Move,* by Darlene Stille. • Introduce wave properties.	*Lesson 1* *Let's Explore Waves!* • Conduct Waves in Motion investigation.	*Lesson 1* *Let's Explore Waves!* • Measure wave amplitudes and wavelengths and make graphs. • Conduct interactive read-aloud of *The Great Wave: A Children's Book Inspired by Hokusai,* by Veronique Massenot. • Create a class list of adjectives to describe *The Great Wave* picture.	*Lesson 1* *Let's Explore Waves!* • Discuss cultural implications of living near an ocean. • Create bar graph of students' wave measurements. • Conduct lesson assessment.

Table 3.7. STEM Road Map Module Schedule for Week Two

Day 6	Day 7	Day 8	Day 9	Day 10
Lesson 2 *Sounds Like Fun!* • Introduce sound and hearing. • Conduct interactive read-aloud of *What Is Sound?* by Charlotte Guillain. • Observe sound waves using plastic forks.	*Lesson 2* *Sounds Like Fun!* • Students describe sounds using adjectives. • Discuss how sound is used for communication. • Introduce hearing disabilities. • Conduct an interactive read-aloud of *The Sense of Hearing,* by Mari Schuh.	*Lesson 2* *Sounds Like Fun!* • Conduct interactive read-aloud of *Sounds All Around,* by Wendy Pfeffer. • Introduce pitch. • Begin What's the Buzz? Investigation.	*Lesson 2* *Sounds Like Fun!* • Complete What's the Buzz? investigation. • Conduct interactive read-aloud of *A Birthday for Ben,* by Kate Gaynor.	*Lesson 2* *Sounds Like Fun!* • Introduce sound measurements (decibels) and estimates. • Introduce the I Can Hear the Music activity and the engineering design process. • Create class list of collaboration rules.

Table 3.8. STEM Road Map Module Schedule for Week Three

Day 11	Day 12	Day 13	Day 14	Day 15
Lesson 2 *Sounds Like Fun!* • Begin the I Can Hear the Music design activity.	*Lesson 2* *Sounds Like Fun!* • Complete the I Can Hear the Music design activity. • Discuss ways to experience sound waves without hearing through discussion and video.	*Lesson 2* *Sounds Like Fun!* • Introduce American Sign Language. • Conduct lesson assessment.	*Lesson 3* *Lighting It Up!* • Introduce light and sight. • Conduct interactive read-aloud of *All About Light,* by Lisa Trumbauer.	*Lesson 3* *Lighting It Up!* • Introduce the concept of the speed of light. • Conduct an interactive read-aloud of *Day Light, Night Light,* by Franklyn M. Branley. • Introduce vision disabilities.

Table 3.9. STEM Road Map Module Schedule for Week Four

Day 16	Day 17	Day 18	Day 19	Day 20
Lesson 3 *Lighting It Up!* • Introduce Lighting the Way investigation. • Conduct interactive read-aloud of *Thomas Edison and His Bright Idea*, by Patricia Brennan Demuth • Conduct lesson assessment.	*Lesson 3* *Lighting It Up!* • Conduct Lighting the Way investigation. • Introduce braille. • Conduct an interactive read-aloud of *Six Dots: A Story of Young Louis Braille*, by Jen Bryant.	*Lesson 3* *Lighting It Up!* • Conduct Shady Shadow Puppets activity. • Students create shadow pictures. • Conduct interactive read-aloud of *My Shadow*, by Robert Louis Stevenson.	*Lesson 3* *Lighting It Up!* • Students present puppet shows. • Introduce eye anatomy and vision with an interactive read-aloud of *Eye: How It Works*, by David Macaulay.	*Lesson 3* *Lighting It Up!* • Conduct interactive read-aloud of *Light Is All Around Us*, by Wendy Pfeffer. • Conduct an interactive read-aloud of *The Black Book of Colors*, by Menena Cottin. • Students create partner description pictures.

Table 3.10. STEM Road Map Module Schedule for Week Five

Day 21	Day 22	Day 23	Day 24	Day 25
Lesson 3 *Lighting It Up!* • Conduct interactive read-aloud of *Sending Messages With Light and Sound*, by Jennifer Boothroyd. • Discuss achievements of person with vision impairment. • Conduct lesson assessment.	*Lesson 4* *Show Me the Waves Challenge* • Discuss use of sound and light in musical shows. • Conduct interactive read-alouds of *I Know a Shy Fellow Who Swallowed a Cello*, by Barbara S. Garriel, and *A Picture Book of Helen Keller*, by David A. Adler. • Introduce Singing Strings activity.	*Lesson 4* *Show Me the Waves Challenge* • Finish Singing Strings activity. • Introduce Incredible Instruments activity. • Conduct interactive read-aloud of *Stand in My Shoes: Kids Learning About Empathy*, by Bob Sornson.	*Lesson 4* *Show Me the Waves Challenge* • Complete Incredible Instruments activity. • Students practice for musical performance. • Teams create poster with team name.	*Lesson 4* *Show Me the Waves Challenge* • Students do their musical performance. • Conduct lesson assessment.

RESOURCES

The media specialist can help you locate resources for students to view and read about waves, sound, light, and related physics content. Special educators and reading specialists can help find supplemental sources for students needing extra support in reading and writing. Additional resources may be found online. Community resources for this module may include musicians, audiologists, optometrists, and mechanical engineers.

REFERENCES

Keeley, P., and R. Harrington. 2010. *Uncovering student ideas in physical science, volume 1: 45 new force and motion assessment probes.* Arlington, VA: NSTA Press.

Koehler, C., M. A. Bloom, and A. R. Milner. 2015. The STEM Road Map for grades K–2. In *STEM Road Map: A framework for integrated STEM education,* ed. C. C. Johnson, E. E. Peters-Burton, and T. J. Moore, 41–67. New York: Routledge. *www.routledge.com/products/9781138804234.*

National Research Council (NRC). 1997. *Science teaching reconsidered: A handbook.* Washington, DC: National Academies Press.

WIDA. 2012. 2012 amplification of the English language development standards: Kindergarten–grade 12. *https://wida.wisc.edu/teach/standards/eld.*

INFLUENCE OF WAVES LESSON PLANS

Andrea R. Milner, Vanessa B. Morrison, Janet B. Walton, Carla C. Johnson, and Erin Peters-Burton

Lesson Plan 1: Let's Explore Waves!

This lesson introduces waves as traveling disturbances that move through space and substances to transfer energy. Students explore the concept of waves through interactive read-alouds, a video, and an inquiry activity in which students explore waves in water.

ESSENTIAL QUESTIONS

- What are waves?

- What kinds of waves are there?

ESTABLISHED GOALS AND OBJECTIVES

At the conclusion of this lesson, students will be able to do the following:

- Explain that waves transfer energy

- Demonstrate how waves in water move

- Identify properties of waves

TIME REQUIRED

- 5 days (approximately 30 minutes each day; see Table 3.6, p. 37)

MATERIALS

Required Materials for Lesson 1

- STEM Research Notebooks (1 per student, see p. 24 for STEM Research Notebook information)

- Computer with internet access for viewing videos
- 1 balloon (nonlatex), inflated
- Radio with speaker
- Books
  - *The Great Wave: A Children's Book Inspired by Hokusai,* by Veronique Massenot (Prestel, 2011).
  - *Waves: Energy on the Move,* by Darlene Stille (Compass Point Books, 2005; we recommend reading Chapter 1, "What Is a Wave?" in this lesson)
- Chart paper
- U.S. map with map scale
- Globe
- Markers
- 10 1-inch × 1-inch building blocks or counting cubes per pair of students
- Plain white paper (1 sheet per student)
- Crayons (1 set of 8 or more per student)
- Safety goggles and apron (per student)

Additional Materials for the Waves in Motion Investigation (per team of 3 or 4 students)

- 1 tub for water (dishpans or large plastic containers)
- About 1 gallon of water
- 1 gallon jug (for students to measure out the water)
- 1 cork
- 5 paper towels

SAFETY NOTES

1. All students should wear safety goggles and aprons during the setup, hands-on, and takedown segments of the activity.

2. Immediately wipe up any water splashed on the floor to avoid a slip-and-fall hazard.

3. Remind students that they should not drink the water used in the activity.

4. Have students wash their hands with soap and water after the activity is completed.

CONTENT STANDARDS AND KEY VOCABULARY

Table 4.1 lists the content standards from the *Next Generation Science Standards* (*NGSS*), *Common Core State Standards* (*CCSS*), National Association for the Education of Young Children (NAEYC), and the Framework for 21st Century Learning that this lesson addresses, and Table 4.2 (p. 46) presents the key vocabulary. Vocabulary terms are provided for both teacher and student use. Teachers may choose to introduce some or all of the terms to students.

Table 4.1. Content Standards Addressed in STEM Road Map Module Lesson 1

NEXT GENERATION SCIENCE STANDARDS

PERFORMANCE EXPECTATION

- 1-PS4-1. Plan and conduct investigations to provide evidence that vibrating materials can make sound and that sound can make materials vibrate.

SCIENCE AND ENGINEERING PRACTICE

Planning and Carrying Out Investigations

Planning and carrying out investigations to answer questions or test solutions to problems in K–2 builds on prior experiences and progresses to simple investigations, based on fair tests, which provide data to support explanations or design solutions.
- Plan and conduct investigations collaboratively to produce evidence to answer a question.

DISCIPLINARY CORE IDEA

PS4.A: Wave Properties

- Sound can make matter vibrate, and vibrating matter can make sound.

CROSSCUTTING CONCEPT

Cause and Effect

- Simple tests can be designed to gather evidence to support or refute student ideas about causes.

Continued

Table 4.1. (*continued*)

COMMON CORE STATE STANDARDS FOR MATHEMATICS

MATHEMATICAL PRACTICES

- MP1. Make sense of problems and persevere in solving them.
- MP2. Reason abstractly and quantitatively.
- MP3. Construct viable arguments and critique the reasoning of others.
- MP4. Model with mathematics.
- MP5. Use appropriate tools strategically.
- MP6. Attend to precision.
- MP7. Look for and make use of structure.
- MP8. Look for and express regularity in repeated reasoning.

MATHEMATICAL CONTENT

- 1.NBT.B.3. Compare two two-digit numbers based on meanings of the tens and ones digits, recording the results of comparisons with the symbols >, =, and <.
- 1.MD.C.4. Organize, represent, and interpret data with up to three categories; ask and answer questions about the total number of data points, how many in each category, and how many more or less are in one category than in another.

COMMON CORE STATE STANDARDS FOR ENGLISH LANGUAGE ARTS

READING STANDARDS

- RI.1.1. Ask and answer questions about key details in a text.
- RI.1.2. Identify the main topic and retell key details of a text.
- RI.1.3. Describe the connection between two individuals, events, ideas, or pieces of information in a text.
- RI.1.7. Use the illustrations and details in a text to describe its key ideas.

WRITING STANDARDS

- W.1.2. Write informative/explanatory texts in which they name a topic, supply some facts about the topic, and provide some sense of closure.
- W.1.6. With guidance and support from adults, use a variety of digital tools to produce and publish writing, including in collaboration with peers.
- W.1.7. Participate in shared research and writing.
- W.1.8. With guidance and support from adults, recall information from experiences or gather information from provided sources to answer a question.

Continued

Table 4.1. (*continued*)

SPEAKING AND LISTENING STANDARDS

- SL.1.1. Participate in collaborative conversations with diverse partners about *grade 1 topics and texts* with peers and adults in small and larger groups.

- SL.1.1.A. Follow agreed-upon rules for discussions.

- SL.1.1.B. Build on others' talk in conversations by responding to the comments of others through multiple exchanges.

- SL.1.1.C. Ask questions to clear up any confusion about the topics and texts under discussion.

- SL.1.3. Ask and answer questions about what a speaker says in order to gather additional information or clarify something that is not understood.

- SL.1.5. Add drawings or other visual displays to descriptions when appropriate to clarify ideas, thoughts, and feelings.

NATIONAL ASSOCIATION FOR THE EDUCATION OF YOUNG CHILDREN STANDARDS

- 2.E.1. Arrange firsthand, meaningful experiences that are intellectually and creatively stimulating, invite exploration and investigation, and engage children's active, sustained involvement by providing a rich variety of material, challenges, and ideas.

- 2.F.3. Extend the range of children's interests and the scope of their thought, present novel experiences and introduce stimulating ideas, problems, experiences, or hypotheses.

- 2.F.6. Enhance children's conceptual understanding through various strategies, including intensive interview and conversation, encourage children to reflect on and "revisit" their experiences.

- 2.G.2. Scaffolding takes on a variety of forms.

- 2.J.1. Incorporate a wide variety of experiences, materials and equipment, and teaching strategies to accommodate the range of children's individual differences in development, skills and abilities, prior experiences, needs, and interests.

- 3.A.1. Teachers consider what children should know, understand, and be able to do across the domains.

FRAMEWORK FOR 21ST CENTURY LEARNING

- Interdisciplinary Themes; Learning and Innovation Skills; Information, Media, and Technology Skills; Life and Career Skills

---

Table 4.2. Key Vocabulary in Lesson 1

Key Vocabulary	Definition
adjective	a word that describes a person, place, or thing
amplitude	the measurement that indicates the strength of a wave; on a graph, it is calculated by finding the height of the wave from the resting position
crest	the highest point of a wave
diffraction	when waves bend and move around obstacles
energy	what causes objects to change and move
frequency	the number of waves that pass a point in a certain amount of time
photon	a bundle of energy that is the basic unit of light
reflection	when waves bounce off an object
refraction	when waves bend as they move from one type of substance to another
speed	how fast something moves
trough	the lowest point of a wave
vibration	movements back and forth
wave height	the distance between the crest and the trough of a wave
wavelength	the distance between wave crests
waves	rolling movements that travel through a substance and transfer energy from one place to another

TEACHER BACKGROUND INFORMATION

First graders are able to make connections across multiple content areas (STEM and English language arts [ELA]) as well as the various developmental domains (physical, social and emotional, personality, cognitive, and language). Incorporating students' prior knowledge with developmentally appropriate instruction will enable them to make these connections. Throughout this module, you should support and facilitate the advancement of these content areas and developmental domains within each student. For information about how formative assessments can be used to connect student prior experiences with classroom instruction, see the STEM Teaching Tools resource "Making Science Instruction Compelling for All Students: Using Cultural Formative Assessment to Build on Learner Interest and Experience" at *http://stemteachingtools.org/pd/sessionc*.

Energy

The standard definition of energy as the ability to do work is an abstract concept that may be difficult for first-grade students to understand. Early elementary students can more easily understand energy from its effects, and therefore energy is conceptualized within the module as how things change and move. There are six types of energy—sound, radiant, mechanical, electrical, chemical, and atomic. This module focuses on sound energy as an example of a type of energy that moves in waves. Connections are also made to light energy within the module, although the wave nature of light is a more abstract concept that may be difficult for young students to conceptualize.

Waves

Waves are a natural part of our daily lives. They are key to communication technologies such as cell phones and radios. Also, we use sunscreen because ultraviolet waves transmit the sun's energy to our skin, and we cook food with microwaves. This module focuses primarily on sound waves with additional connections to light waves. The movement of sound and light in waves is an abstract concept for students, and therefore this lesson gives students the opportunity to learn about the properties of waves by investigating waves that they can see and touch—waves in water.

This lesson focuses on wind as a cause of waves in water. Students will investigate waves in water to understand that waves transfer energy from one place to another, and that waves have various characteristics that can vary depending on the amount of energy they carry. The following websites provide additional background information about ocean waves:

- *www.oceanservice.noaa.gov/facts/wavesinocean.html*

- *www.nationalgeographic.org/media/explaining-waves*

Waves are typically described by their properties. These properties include amplitude, wavelength, frequency, and speed. Waves are often depicted graphically to demonstrate their properties. Amplitude is the strength of the wave, or its displacement of the vibrating matter from its rest position (measured as its height from its resting position). Wave height is the distance between the crest and trough of a wave. Wavelength is the distance between two points of the wave (e.g., the distance between crests or troughs of a wave). The frequency of a wave is the number of cycles a wave goes through in a given period of time, and the speed or velocity of the wave is how fast the wave travels through a given medium. A wave's velocity can be found by multiplying its frequency by its wavelength. More information about wave properties can be found at *www.ducksters. com/science/physics/properties_of_waves.php*.

You may wish to introduce careers associated with waves and communication technologies during this module, such as the following (adapted from Koehler, Bloom, and Milner 2015):

- audio engineer

- electrician

- hearing specialist

- light engineer

- musician

- optometrist

For more information about these and other careers, see the Bureau of Labor Statistics' *Occupational Outlook Handbook* at *www.bls.gov/ooh/home.htm*.

Sound

Sound is what we hear when sound waves reach our ears. Sound is produced when an object vibrates, producing sound waves. When mechanical energy (e.g., plucking a guitar string) causes an object to vibrate, it causes particles of air to vibrate that then collide with other air particles close to them. The vibration passes from particle to particle, creating waves. Since sound waves are the vibration of particles, sound must move through some type of matter or medium. This means that in a vacuum, or in outer space, there is no sound. The size and shape of the waves reaching our eardrums determines the type of sound we hear. The speed with which sound waves move depends on the medium through which they move. Sound waves collide with particles that make up matter and make those particles vibrate. Sound waves travel when these vibrations are passed from particle to particle, so that sound moves faster in substances with closely spaced particles. For example, sound waves move faster in water and in steel than in air. Sound moves at about 343 meters per second or 768 miles per hour in air (about 1 mile in 5 seconds). The rate of speed is about four times faster in water and about 13 times faster in steel.

For sound waves, the amplitude or height of the wave is in direct proportion to the loudness of the sound. The volume of sound is expressed in decibels, a measure of the intensity of the sound. Sounds louder than about 85 decibels can damage a person's hearing depending on the amount of time of exposure. For comparison purposes, a normal conversation voice is about 60 decibels, a lawnmower creates sound of about 90 decibels, and a rock concert creates sound of about 120 decibels.

Light

Light is a type of electromagnetic radiation that moves in waves visible to our eyes. Light is unique since it behaves both as waves and as particles called photons. The treatment of light in this module will focus on light's observable properties. Light can travel through some substances like air and water (those substances are called transparent), but not through some solid substances such as a wall or a desk (these substances are called opaque). These opaque substances reflect the light entirely. Lesson 3 includes definitions for and further discussions of the terms *transparent* and *opaque*. The speed at which light travels is the fastest known speed. If there is no substance to slow light (e.g., in a vacuum), it moves at 186,282 miles per second. Light is slowed somewhat when it travels through transparent substances such as air and water; however, even through these substances, it travels far faster than we could measure with standard, everyday instruments such as a stopwatch.

Light waves are characterized by certain types of behavior, including reflection, refraction, and diffraction. The light we see is reflected light, and the colors we see are a result of the way that light reflects from objects. When light waves travel from one medium to another (e.g., from air to water), they refract, or change directions. The waves may also slow down or speed up, depending on the substances they are moving between. An example of refraction is looking at a fish in a fish bowl from above. The apparent location of the fish is different than the actual location of the fish due to the change in direction of the light waves. Diffraction occurs when light waves pass from one medium to another and bend. Examples of diffraction are the rainbow pattern we see when looking at a CD or DVD and the ring of light sometimes visible around the Moon (small particles in the atmosphere diffract the light). For more information about reflection, refraction, and diffraction, see *https://science.nasa.gov/ems/03_behaviors*.

We can only see light waves within the visible spectrum, a range of wavelengths and frequencies of light visible by the human eye. Other electromagnetic waves, such as microwaves, radio waves, and x-rays, behave similarly to light waves but have wavelengths and frequencies outside of the visible spectrum and are therefore not visible to humans. When light waves collide with an object, some wavelengths of light are absorbed by the object and some are reflected; it is the wavelength of the reflected waves that determine the color we see. When we see white, the object is equally reflecting all wavelengths; when we see black, the object is absorbing all or nearly all wavelengths of light. For more information about the properties of light waves, see *www.ducksters.com/science/light_spectrum.php*.

ELA Connection: Literature Inspired by Katsushika Hokusai

In this lesson, students will connect their learning about ocean waves to art and literature by considering the descriptive power of words and pictures as illustrated by a

children's book inspired by Katsushika Hokusai's famous woodblock print, *The Great Wave*. Hokusai, who lived and worked in the mid-1800s, was a Japanese painter and woodblock artist. The sea was a subject of many of Hokusai's works, although the focus of the series of prints of which *The Great Wave* is a part was actually Mount Fuji, which is visible in the background. To view Hokusai's work and for more information about the artist, see *www.mymodernmet.com/katsushika-hokusai-the-great-wave/2*.

This famous work inspired children's book author Veronique Massenot to write a children's story using *The Great Wave* as a starting point. This book provides an entry point for students to consider how waves can be represented in art and literature with a focus on the use of descriptive words (adjectives) and art work. Students should understand that adjectives are words that are used to describe people, places, and things. They will be challenged to identify the use of adjectives to describe people, places, and things in literature, to use adjectives to describe images, and to create their own pictures based on a list of adjectives.

Know, Learning, Evidence, Wonder, Scientific Principles (KLEWS) Charts

Throughout this module, you will track student knowledge on Know, Learning, Evidence, Wonder, Scientific Principles (KLEWS) charts. These charts are used to access and assess student prior knowledge, encourage students to think critically about the topic under discussion, and track student learning throughout the module. Using KLEWS charts challenges students to connect evidence and scientific principles with their learning. Be sure to list the topic at the top of each chart. The charts should consist of five columns—one for each KLEWS component. It may be helpful to post these charts in a prominent place in the classroom so that students can refer to them throughout the module. Students will write their personal ideas and reflections in their STEM Research Notebooks entries. For more information about KLEWS charts, see the January 2006 National Science Teaching Association *WebNews Digest* article "Evidence Helps the KWL get a KLEW" at *www.nsta.org/publications/news/story.aspx?id=51519* or the February 2015 *Science and Children* article "Methods and Strategies: KLEWS to Explanation-Building in Science" at *www.nsta.org/store/product_detail.aspx?id=10.2505/4/sc15_052_06_66*.

Interactive Read-Alouds

This module also uses interactive read-alouds to engage students, access their prior knowledge, develop student background knowledge, and introduce topical vocabulary. These read-alouds expose children to teacher-read literature that may be beyond their independent reading levels but is consistent with their listening level. Interactive read-alouds may incorporate a variety of techniques, and you can find helpful information regarding these techniques at the following websites:

- *www.readingrockets.org/article/repeated-interactive-read-alouds-preschool-and-kindergarten*

- *www.k5chalkbox.com/interactive-read-aloud.html*

- *www.readwritethink.org/professional-development/strategy-guides/teacher-read-aloud-that-30799.html*

In general, interactive read-alouds provide opportunities for students to share prior knowledge and experiences, interact with the text and concepts introduced therein, launch conversations about the topics introduced, construct meaning, make predictions, and draw comparisons. You may wish to mark places within the texts to pause to ask for student experiences, predictions, or other ideas. Each reading experience should focus on an ongoing interaction between students and the text, including the following:

- Allow students to share personal stories throughout the reading.

- Ask students to predict throughout the story.

- Allow students to add new ideas from the book to the KLEWS chart and their STEM Research Notebooks.

- Allow students to add new words from the book to the vocabulary chart and their STEM Research Notebooks.

The materials list for each lesson includes the books for interactive read-alouds that you will use in that lesson. A list of suggested books for additional reading can be found at the end of this chapter (see p. 120).

COMMON MISCONCEPTIONS

Students will have various types of prior knowledge about the concepts introduced in this lesson. Table 4.3 (p. 52) outlines some common misconceptions students may have concerning these concepts. Because of the breadth of students' experiences, it is not possible to anticipate every misconception that students may bring as they approach this lesson. Incorrect or inaccurate prior understanding of concepts can influence student learning in the future, however, so it is important to be alert to misconceptions such as those presented in the table.

Table 4.3. Common Misconceptions About the Concepts in Lesson 1

Topic	Student Misconception	Explanation
Waves	When a wave moves through a substance, it carries particles of that substance along with it in the direction the wave is moving.	Waves are disturbances that transport energy from one place to another; although particles of matter vibrate in waves, they do not move along with the wave.
	Waves occur only in water.	We can see water moving in a wave-like motion; however, there are waves around us all the time. For example, sound and light both move in waves.

PREPARATION FOR LESSON 1

Review the Teacher Background Information section (p. 46), assemble the materials for the lesson, duplicate the student handouts, and preview the video recommended in the Learning Components section that follows. Present students with their STEM Research Notebooks and explain how they will be used (see p. 24). Templates for the STEM Research Notebook are provided in Appendix A, and a rubric for observations, student participation, and STEM Research Notebook entries is provided in Appendix B.

Prepare a wave graph on chart paper for each pair of students to use to measure wave amplitudes and wavelengths. Use a horizontal line to indicate calm sea level (see Figure 4.1, p. 61) and draw waves that oscillate above and below this line. The graphs should vary by wave amplitude and wavelength so that student pairs can compare their wave measurements with those of other student pairs. Students will make measurements using 1-inch × 1-inch blocks, so wave amplitudes and wave lengths should be between 1 and 10 inches.

STEM Research Notebook Entry #3 provides a template for students to record vocabulary words. You may wish to use this template throughout the module for students to record definitions and illustrations of key vocabulary words. The template provides space for definitions and illustrations of three words. If you introduce more than three vocabulary words in a lesson, you should make multiple copies of the template for each student.

For the ELA Connection (see Explanation section, p. 57), you should prepare two lists of four to five adjectives that are opposite or very different from each other. Half the class will create pictures based on one list of adjectives and the other half will create pictures based on the second list. These lists may focus on ocean waves to tie into the lesson topic. For example, you may wish to provide lists such as the following:

- Group 1: windy, cloudy, dark, large, crashing

- Group 2: calm, sunny, bright, small, gentle

LEARNING COMPONENTS
Introductory Activity/Engagement

Connection to the Challenge: Begin each day of this lesson by directing students' attention to the module challenge, the Show Me the Waves Challenge:

> *You and your team will be challenged to create a musical show that demonstrates how waves can be used to communicate over distances. You will create musical instruments that create sound waves, and you will also use light to show how sound and light can be used to communicate and entertain.*

Tell students that they will learn about different types of waves, both those that they can see with their eyes and those that they can hear and feel. Hold a brief class discussion on how students' learning in the previous days' lessons contributed to their ability to complete the challenge. You may wish to create a class list of key ideas on chart paper.

Science Class: Introduce the concept of waves with a class discussion. Following agreed-upon rules for discussions, hold a class discussion about waves, asking students the following:

- What are waves?

- Are there different types of waves?

- What kinds of waves are there?

- What makes waves?

- Where and when have you seen waves?

As students share their ideas and prior experiences, chart student responses on a KLEWS chart in the Know column. Then, ask students what questions they have about waves, recording these questions in the Want to Know column.

Introduce students to the STEM Research Notebooks they will use throughout the module. Explain to students that scientists and other STEM workers use research notebooks to track their work.

Then, show a video about ocean waves such as the drone video found at *www.wtkr.com/2014/01/17/this-drone-shot-video-of-surfers-in-hawaii-will-blow-your-mind*. Hold a class discussion about what students observed about waves in this video. After viewing the video, ask students to share what they learned about waves, adding their responses to the KLEWS chart. Students will also record their ideas in STEM Research Notebook Entry #1.

STEM Research Notebook Entry #1

Have students document their own ideas about waves after watching the video in their STEM Research Notebooks, using both words and pictures.

Next, introduce the idea that waves move energy from one place to another. Ask students to share their ideas about what energy is, tracking student ideas on a KLEWS chart. Guide students to understand that energy is anything that causes objects to change and move. Place a balloon on a desk or on the floor where it is visible to the class. Ask students to share their ideas about what could change the balloon or make it move. Students will likely offer examples of mechanical energy (e.g., kicking, throwing). Ask students to share the energy source for each response they offer. Then, ask students what would happen to the balloon if you put it by the window on a sunny day (it would get warm). Ask students what the energy source is in this case (the sun, or light). After that, have a student lightly touch one side of the balloon while you hold the other side to the radio speaker with the radio turned off. Ask the student what he or she feels (nothing). Next, turn the radio on and ask the student what he or she feels (vibrations). Ask students what the energy source is in this case (your voice, or sound). Add student responses to the KLEWS chart.

Then, ask students to think again about waves in water. Ask students if all waves look the same. Ask students to name the ways that waves are different (e.g., some are taller than others). Tell students that all waves have different features called properties and that waves are described by these properties. Introduce wave properties (height or amplitude, wavelength, trough, crest, and frequency) by showing students Figure 4.1 attached at the end of this lesson (see p. 61). Ask students to share their ideas about what factors influence these wave properties in the ocean (e.g., wind).

Conduct an interactive read-aloud of *Waves: Energy on the Move*, by Darlene Stille (specifically, Chapter 1, "What Is a Wave?"). Students will create STEM Research Notebook entries after the read-aloud. You should continue to track student responses on the KLEWS chart.

STEM Research Notebook Entry #2

After the read-aloud, ask students to reflect on what they learned about waves and the causes of waves in their STEM Research Notebooks, using both words and pictures.

Mathematics Connection: Introduce the idea that waves' properties can be measured. Show students Figure 4.1 and ask them to name what properties they could measure. Draw several wave graphs on the board and model measuring the wave amplitudes and wavelengths using the 1-inch blocks.

ELA Connection: Begin a class vocabulary chart using vocabulary from the interactive read-aloud and the class discussion of wave properties. Use pictures to illustrate the vocabulary and post the chart on the classroom wall in a location where it can be easily referenced. You will add to this chart throughout the module.

STEM Research Notebook Entry #3

Have students add vocabulary words to their STEM Research Notebooks, using both words and pictures.

Social Studies Connection: Ask students in what part of the country or the world they think people can surf. Help students locate on the map or globe the regions throughout the United States or the world where people are able to surf. Discuss why those regions are ideal locations for surfing (i.e., waves are large).

Activity/Exploration

Science Class: Students will work in teams of three or four to investigate waves in water in the Waves in Motion investigation, using the Predict, Observe, Explain (POE) process. Students will create STEM Research Notebook entries (specifically, entries 4–6) for each step of this process, and you should also track student predictions, observations, and explanations on a class POE) chart.

Waves in Motion Investigation

In this investigation, students will work in teams of three or four to create waves in tubs of water in which a "boat" (a cork) is floating. Students will disrupt the water in various ways to create waves (e.g., tilting the pan, blowing on the water, splashing the water, clapping, pounding on table). Introduce the activity by showing students the materials they will work with. Hold a class discussion to predict what will happen in the wave pan. Create a POE chart to track student responses.

Tell students that they are going to make predictions about what will happen in their wave pans. Ask students to respond to the following questions in a class discussion and in STEM Research Notebook Entry #4:

- Can we make waves?

- How can we make different types of waves?

- How will different ways of making the water move affect the cork (boat)?

STEM Research Notebook Entry #4

Have students record their predictions about how they can create waves, how they can create different sizes of waves, and what effect these waves will have on their cork boats.

Next, distribute materials (tubs, corks, water, water jugs, and paper towels) to student teams. Have each team fill its tub about one-third to one-half full of water and place the cork in the water. Next, ask students to disrupt the water (create waves) in various ways (e.g., tilting the pan, blowing on the water, splashing the water, clapping, pounding on table). Have students observe the effects of their actions on the waves and the effects of the waves on their cork boats. Have students share their observations in a class discussion and record them in STEM Research Notebook Entry #5, asking students the following:

- How did you make waves?

- How did you make different sizes of waves?

- How did different sizes of waves affect the cork (boat)?

STEM Research Notebook Entry #5

Have students record their observations about how they made waves and what effect those waves had on their cork boats in their STEM Research Notebooks.

Mathematics Connection: Group students in pairs and distribute the wave graphs you created to each student pair (see Preparation for Lesson 1, p. 52). Review with students how to measure the wave's height and wavelength using the 1-inch blocks and count the number of crests and troughs on their wave. Have student pairs work together to measure their waves and record their findings in STEM Research Notebook Entry #6. After student pairs have measured their waves and recorded their findings, have each pair work with another pair of students to compare their wave amplitudes, wavelengths, and numbers of crests and troughs and enter this information in the STEM Research Notebook entry.

STEM Research Notebook Entry #6

Have students record the wave amplitudes and wavelengths they measured, the number of crests and troughs they counted, and how their waves compare with another student pair's waves in their STEM Research Notebooks.

ELA Connection: Ask students to name some words that describe water and waves, creating a class list. Introduce the idea that words that describe a person, place, or object are called adjectives. Tell students that they are going to read aloud a book as a class and that they should raise their hands each time they hear a describing word, or adjective. Conduct an interactive read-aloud of the book *The Great Wave: A Children's Book Inspired by Hokusai,* by Veronique Massenot. Create a class list of adjectives that are used to describe the wave in this book. Tell students that *The Great Wave* is actually a famous work of art by a Japanese artist named Hokusai, and that the author of the book used this work of art to inspire a story. Ask students to share their ideas about how a picture can inspire a story and what elements of Hokusai's work may have inspired the author of the book to write about it.

Next, show students Figure 4.2, the image of a dog surfing attached at the end of the lesson (see p. 61), and have students discuss what adjectives describe the picture and complete STEM Research Notebook Entry #7.

STEM Research Notebook Entry #7

Have students look at the image of the surfing dog and create a list of adjectives that describe the picture.

Social Studies Connection: Not applicable.

Explanation

Science Class: As a class, discuss what students observed in the Waves in Motion investigation. As a class, also revisit predictions for the Waves in Motion investigation. Have students compare their predictions to their observations. Ask students to explain the causes for what happened. Students will discuss whether their predictions were accurate, close, or not accurate and why. Document student responses on the POE chart. Next, ask students to explain what caused waves in their wave tubs. Have students complete STEM Research Notebook Entry #8 to explain cause and effect for their observations in the Waves in Motion investigation.

STEM Research Notebook Entry #8

Have students record their explanations for their observations by completing a cause-and-effect chart in their STEM Research Notebooks, using both words and pictures.

Mathematics Connection: Create a table on the board or on chart paper with columns for students to record their pair's wave amplitudes and wavelengths and numbers of peaks and troughs. Have students identify both the smallest wave amplitude and the largest wave amplitude on the chart. Have the student pairs with the corresponding wave graphs display their graphs. Next, have students identify the smallest wavelength and the largest wavelength on the chart. Have the student pairs with the corresponding wave graphs display their graphs.

ELA Connection: Remind students that both words and pictures can describe. Split the class into two groups, and provide each group with a list of adjectives (see Preparation for Lesson 1, p. 52) that are opposite or very different from the other group's adjectives.

Provide each student with a piece of plain white paper and crayons, and have each student each create a picture that reflects the adjectives for their group.

Social Studies Connection: Revisit the map or globe locations identified during the Introductory Activity/Engagement. Hold a class discussion about the cultural implication of living near an ocean, asking the following:

- What kind of sports can you play if you live near an ocean that you cannot play as easily if you live far from the ocean?

- Would we find more surfers living in regions by oceans or living in inland regions? Why?

- How else would life be the same or different living near the ocean compared with living far from the ocean?

- What are the advantages of living near the ocean?

- What are the disadvantages of living near the ocean?

Elaboration/Application of Knowledge

Science Class: Assess student learning for the lesson using the assessment instrument provided in Appendix B (p. 162).

Mathematics Connection: Introduce the idea of bar graphs as a way to visually organize and display numerical information. As a class, work together to create a bar graph of the wave amplitudes students measured using the table the class completed in the Explanation section.

ELA Connection: After students have completed their pictures (see Explanation section), pair each student with a student from the other group. Have students compare and contrast their pictures, identifying similarities and differences. Next, have student pairs share their ideas with the class and hold a class discussion about how the adjectives provided resulted in different types of pictures.

Social Studies Connection: Using a U.S. map, have students identify their location and the location of the nearest ocean. Demonstrate to students how to use a map scale to determine distances on a map and work together as a class to find the approximate number of miles to the nearest ocean. If you are located in a coastal area, you may find the approximate number of miles to the opposite coast.

Evaluation/Assessment

Students may be assessed on the following performance tasks and other measures listed.

Performance Tasks

- Waves in Motion investigation

- Wave measurements

- Lesson 1 Assessment (see Appendix B, p. 162)

Other Measures (see assessment rubric in Appendix B, p. 170)

- Teacher observations

- STEM Research Notebook entries

- Participation in teams during investigations

INTERNET RESOURCES

Formative assessment
- *http://stemteachingtools.org/pd/sessionc*

Ocean waves
- *www.oceanservice.noaa.gov/facts/wavesinocean.html*

- *www.nationalgeographic.org/media/explaining-waves*

Wave properties
- *www.ducksters.com/science/physics/properties_of_waves.php*

Bureau of Labor Statistics' *Occupational Outlook Handbook*
- *www.bls.gov/ooh/home.htm*

Behavior and properties of light waves
- *https://science.nasa.gov/ems/03_behaviors*

- *www.ducksters.com/science/light_spectrum.php*

Katsushika Hokusai's *The Great Wave*
- *www.mymodernmet.com/katsushika-hokusai-the-great-wave/2*

KLEWS charts
- *www.nsta.org/publications/news/story.aspx?id=51519*

- *www.nsta.org/store/product_detail.aspx?id=10.2505/4/sc15_052_06_66*

Interactive read-alouds

- *www.readingrockets.org/article/repeated-interactive-read-alouds-preschool-and-kindergarten*

- *www.k5chalkbox.com/interactive-read-aloud.html*

- *www.readwritethink.org/professional-development/strategy-guides/teacher-read-aloud-that-30799.html*

Video of ocean waves

- *www.wtkr.com/2014/01/17/this-drone-shot-video-of-surfers-in-hawaii-will-blow-your-mind*

Figure 4.1. Wave Properties

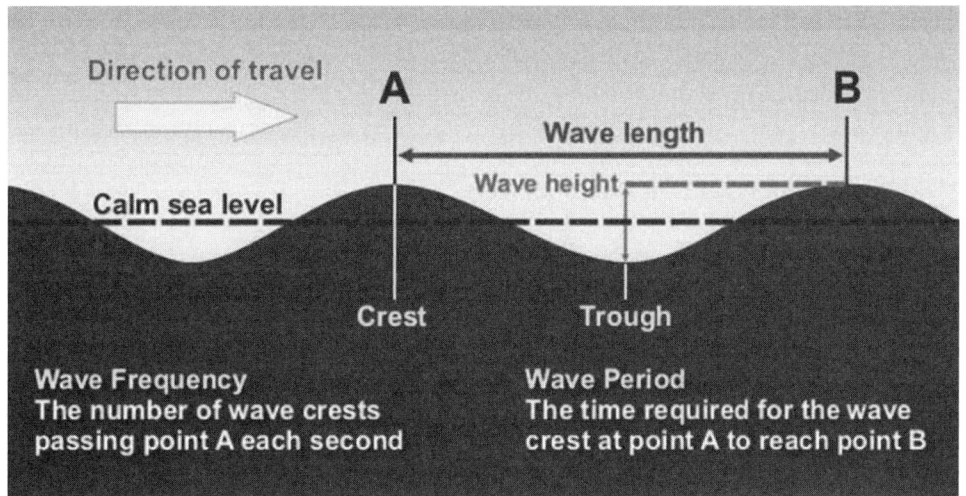

Source: Retrieved from the National Oceanic and Atmospheric Administration at *www.oceanservice.noaa.gov/education/kits/currents/media/supp_cur03a.html.*

Figure 4.2. Image of Dog Surfing

Note: A full-color version of this image is available on the book's Extras page at *www.nsta.org/roadmap-waves.*

Influence of Waves, Grade 1

Lesson Plan 2: Sounds Like Fun!

In this lesson, students explore the concept of sound through interactive read-alouds, videos, and inquiry activities. Students create a simple communication device to investigate the role of sound in communicating over distances. Students also learn the steps of the engineering design process (EDP) and work in teams to create a musical instrument.

ESSENTIAL QUESTIONS

- What is sound?

- How does sound travel over distances?

- How do we hear?

- How do we use sound to communicate?

ESTABLISHED GOALS AND OBJECTIVES

At the conclusion of this lesson, students will be able to do the following:

- Understand that there are various sources of sound waves and identify sources of sound

- Explain how sound waves travel to reach human ears

- Demonstrate a basic understanding of the human anatomy associated with hearing

- Design, construct, test, and evaluate models that demonstrate how humans experience and interact with sound waves

- Explain that sound can be used to communicate over distances

- Evaluate the influence sound waves have on culture and society

TIME REQUIRED

- 8 days (approximately 30 minutes each day; See Tables 3.7–3.8, p. 38)

MATERIALS

Required Materials for Lesson 2

- STEM Research Notebooks

- Computer with internet access for viewing videos

- Books

  - *A Birthday for Ben,* by Kate Gaynor (Special Stories Publishing, 2009)

  - *Sounds All Around,* by Wendy Pfeffer (HarperCollins, 2016)

  - *The Sense of Hearing,* by Mari Schuh (Bellwether Media, 2007)

  - *What Is Sound?* by Charlotte Guillain (Heinemann, 2009)

- Chart paper

- Markers

- Medium or heavy weight disposable plastic forks (1 per student, see Preparation for Lesson 2, p. 72)

- Nail or sharp tool to punch holes in bottom of plastic cups

- 2 round, clear plastic containers, about a 3-cup capacity

- Medium- or large-sized beads or marbles (about ½ inch in diameter) sufficient to fill a 3-cup container to the top with an additional 10 beads or marbles left over

- Radio with speaker

- Safety glasses with side shields or safety goggles (per student)

Additional Materials for What's the Buzz? (1 set per pair of students)

- 2 8- to 12-ounce plastic cups with pre-punched holes in bottom

- 3 pieces of cotton string in the following measurements:

  - 5 feet (60 inches)

  - 10 feet (120 inches)

  - 15 feet (180 inches)

- Scissors

- Yardstick

Additional Materials for I Can Hear the Music (1 set per team of 3 or 4 students unless otherwise noted)

- 2 or 3 plastic, metal, or cardboard containers (e.g., oatmeal box, small plastic container, aluminum can) of varying sizes

- 2 linear feet of plastic wrap

- 2 linear feet of waxed paper
- 2 linear feet of aluminum foil
- 4 scraps of fabric (large enough to cover openings of containers)
- 1 paper plate
- 8 rubber bands (nonlatex)
- ½ cup of rice
- ½ cup of dry beans
- 2 plastic spoons
- 5 large craft sticks
- 1 cork
- Masking tape
- Scissors
- 1 sheet of plain white paper (per student)

SAFETY NOTES

1. Remind students that safety glasses with side shields or safety goggles must be worn during all phases of this inquiry activity.

2. Immediately pick up any items dropped on the floor to avoid a slip or trip-and-fall hazard.

3. Students should use caution when working with sharps (e.g., scissors, nail, stick, can) as they can cut or puncture skin.

4. Students should have direct adult supervision if they are puncturing a hole in bottom of a container.

5. Remind students to never stick pointed objects in their ears because they can rupture eardrums.

6. Remind students to never put beads in their mouth because of the potential choking issue.

7. Have students wash their hands with soap and water after the activity is completed.

CONTENT STANDARDS AND KEY VOCABULARY

Table 4.4 lists the content standards from the *NGSS, CCSS,* NAEYC, and the Framework for 21st Century Learning that this lesson addresses, and Table 4.5 (p. 68) presents the key vocabulary. Vocabulary terms are provided for both teacher and student use. Teachers may choose to introduce some or all of the terms to students.

Table 4.4. Content Standards Addressed in STEM Road Map Module Lesson 2

NEXT GENERATION SCIENCE STANDARDS

PERFORMANCE EXPECTATIONS

- 1-PS4-1. Plan and conduct investigations to provide evidence that vibrating materials can make sound and that sound can make materials vibrate.

- 1-PS4-4. Use tools and materials to design and build a device that uses light or sound to solve the problem of communicating over a distance.

SCIENCE AND ENGINEERING PRACTICES

Planning and Carrying Out Investigations

Planning and carrying out investigations to answer questions or test solutions to problems in K–2 builds on prior experiences and progresses to simple investigations, based on fair tests, which provide data to support explanations or design solutions.
- Plan and conduct investigations collaboratively to produce evidence to answer a question.

Constructing Explanations and Designing Solutions

Constructing explanations and designing solutions in K–2 builds on prior experiences and progresses to the use of evidence and ideas in constructing evidence-based accounts of natural phenomena and designing solutions.
- Make observations (firsthand or from media) to construct an evidence-based account for natural phenomena.

DISCIPLINARY CORE IDEAS

PS4.A: Wave Properties
- Sound can make matter vibrate, and vibrating matter can make sound.

PS4.C : Information Technologies and Instrumentation
- People also use a variety of devices to communicate (send and receive information) over long distances.

Continued

Table 4.4. (*continued*)

CROSSCUTTING CONCEPT

Cause and Effect

- Simple tests can be designed to gather evidence to support or refute student ideas about causes.

COMMON CORE STATE STANDARDS FOR MATHEMATICS

MATHEMATICAL PRACTICES

- MP1. Make sense of problems and persevere in solving them.
- MP2. Reason abstractly and quantitatively.
- MP3. Construct viable arguments and critique the reasoning of others.
- MP4. Model with mathematics.
- MP5. Use appropriate tools strategically.
- MP6. Attend to precision.
- MP7. Look for and make use of structure.
- MP8. Look for and express regularity in repeated reasoning.

MATHEMATICAL CONTENT

- 1.NBT.A.1. Count to 120, starting at any number less than 120. In this range, read and write numerals and represent a number of objects with a written numeral.
- 1.NBT.B.3. Compare two two-digit numbers based on meanings of the tens and ones digits, recording the results of comparisons with the symbols >, =, and <.

COMMON CORE STATE STANDARDS FOR ENGLISH LANGUAGE ARTS

READING STANDARDS

- RI.1.1. Ask and answer questions about key details in a text.
- RI.1.2. Identify the main topic and retell key details of a text.
- RI.1.3. Describe the connection between two individuals, events, ideas, or pieces of information in a text.
- RI.1.7. Use the illustrations and details in a text to describe its key ideas.

WRITING STANDARDS

- W.1.2. Write informative/explanatory texts in which they name a topic, supply some facts about the topic, and provide some sense of closure.
- W.1.6. With guidance and support from adults, use a variety of digital tools to produce and publish writing, including in collaboration with peers.
- W.1.7. Participate in shared research and writing.

Continued

Table 4.4. (*continued*)

- W.1.8. With guidance and support from adults, recall information from experiences or gather information from provided sources to answer a question.

SPEAKING AND LISTENING STANDARDS

- SL.1.1. Participate in collaborative conversations with diverse partners about *grade 1 topics and texts* with peers and adults in small and larger groups.

- SL.1.1.A. Follow agreed-upon rules for discussions.

- SL.1.1.B. Build on others' talk in conversations by responding to the comments of others through multiple exchanges.

- SL.1.1.C. Ask questions to clear up any confusion about the topics and texts under discussion.

- SL.1.3. Ask and answer questions about what a speaker says in order to gather additional information or clarify something that is not understood.

- SL.1.5. Add drawings or other visual displays to descriptions when appropriate to clarify ideas, thoughts, and feelings.

NATIONAL ASSOCIATION FOR THE EDUCATION OF YOUNG CHILDREN STANDARDS

- 2.E.1. Arrange firsthand, meaningful experiences that are intellectually and creatively stimulating, invite exploration and investigation, and engage children's active, sustained involvement by providing a rich variety of material, challenges, and ideas.

- 2.F.3. Extend the range of children's interests and the scope of their thought, present novel experiences and introduce stimulating ideas, problems, experiences, or hypotheses.

- 2.F.6. Enhance children's conceptual understanding through various strategies, including intensive interview and conversation, encourage children to reflect on and "revisit" their experiences.

- 2.G.2. Scaffolding takes on a variety of forms.

- 2.J.1. Incorporate a wide variety of experiences, materials and equipment, and teaching strategies to accommodate the range of children's individual differences in development, skills and abilities, prior experiences, needs, and interests.

- 3.A.1. Teachers consider what children should know, understand, and be able to do across the domains.

FRAMEWORK FOR 21ST CENTURY LEARNING

- Interdisciplinary Themes; Learning and Innovation Skills; Information, Media, and Technology Skills; Life and Career Skills

Table 4.5. Key Vocabulary in Lesson 2

Key Vocabulary	Definition
cochlea	a part of the inner ear that is filled with fluid and nerve cells that respond to specific frequencies of sound
decibels	the unit used to measure how loud a sound is
eardrum	a thin, tightly stretched flap of skin in the middle ear that vibrates when sound waves hit it
estimate	an approximate value that is based on incomplete information
hearing	the process of the ear receiving sound waves and sending messages to the brain
inner ear	the part of the ear where the cochlea is located
medium	a substance through which energy can move
middle ear	the part of the ear where the eardrum is located
nerve	the parts of the body that carry signals to and from the brain
outer ear	the portion of the ear that is visible and that directs sound waves toward the eardrum
pitch	the highness or lowness of a sound
propagate	to spread
sound	what we hear when sound waves enter our ears
sound wave	vibrating energy that travels from one place to another; when sound waves enter our ears, we hear sound
volume	a measure of loudness

TEACHER BACKGROUND INFORMATION

This lesson focuses on sound waves, how sound waves can be used to communicate over distances, and how the human body responds to sound waves.

Sound and Human Hearing

When sound waves travel through a medium, the particles of that medium vibrate with a specific frequency (a specific number of waves in a given amount of time). These frequencies, measured in units called Hertz (Hz) determine whether the sound can be perceived by humans. One Hertz is equivalent to one vibration per second. The human ear can detect sound waves in a wide range of frequencies (about 20 Hz to 20,000 Hz) and sends electrical signals to the brain which then interprets that sound. Different species of animals can hear different ranges of frequencies. Dogs' ears, for example, can detect frequencies from about 50 Hz to 45,000 Hz, and bats can detect frequencies of up to 120,000 Hz. The brain interprets the specific frequency of a sound wave as pitch, with a high frequency sound wave interpreted as a high pitch and a low frequency sound wave as a low pitch. In this lesson, students will relate wavelength to pitch; sound waves with longer wavelengths have lower pitches and those with shorter wavelengths have higher pitches. For more information about sound waves' pitch and frequencies see *www.physicsclassroom.com/class/sound/Lesson-2/Pitch-and-Frequency*.

Human hearing is a complex combination of physics, anatomy, and psychology. This lesson's content focuses on the basic physics and anatomy of hearing, in which sound waves are converted to mechanical energy in the human ear, which in turn sends nerve impulses to the brain to interpret that energy as sound. The human ear has three essential parts: the outer ear, the middle ear, and the inner ear (see Figure 4.3, p. 85). These parts each play a role in how we receive and process sound waves. The outer ear directs sound waves to the middle ear where the sound waves create vibrations in the eardrum and the three bones of the middle ear: the malleus, incus, and stapes (commonly called the hammer, anvil, and stirrup). The cochlea in the inner ear contains fluid and many nerve cells. The nerve cells are each sensitive to a specific frequency of vibration. As vibrations from the middle ear pass through this fluid, they stimulate the nerve cells that are sensitive to the vibration's frequency to release an electrical signal that the auditory nerve (the cochlear nerve) carries to the brain. The brain then interprets these nerve signals as specific sounds. For more information about human ear anatomy and the science of hearing, see the following websites:

- *www.physicsclassroom.com/class/sound/Lesson-2/The-Human-Ear*

- *www.asha.org/public/hearing/How-We-Hear*

The complex mechanisms associated with hearing mean that there are many different types of and causes of hearing loss. Damage to any component of the human anatomy associated with receiving and processing sound waves can result in hearing loss. For example, hearing loss can be a result of age-related changes in the middle ear and neural pathways that transmit signals to the brain, exposure to loud noises (either sudden loud noises such as explosions or sustained exposure to loud sounds over time) that damage the nerve cells in the cochlea, middle ear infections that cause temporary hearing loss, and auditory processing disorders in which the brain has difficulty interpreting the electrical impulses associated with sound waves. For more information on the types of and causes of hearing loss, see the National Institutes of Health "Hearing, Ear Infections, and Deafness" page at *www.nidcd.nih.gov/health/hearing-ear-infections-deafness*.

Engineering

Students begin to gain an understanding of engineering as a profession in this lesson as they learn to use the EDP to create musical instruments. Students should understand that engineers are people who design and build products and systems in response to human needs. For an overview of the various types of engineering professions, see the following websites:

- *www.engineergirl.org/33/TryOnACareer*

- *www.nacme.org/types-of-engineering*

- *www.sciencekids.co.nz/sciencefacts/engineering/typesofengineeringjobs.html*

In this lesson, students will work in teams to create musical instruments that vibrate like a human eardrum and incorporate other musical features (e.g., rattle when shaken). Students will work with simple sets of materials to create two types of musical instruments.

Engineering Design Process

Students should understand that engineers need to work in groups to accomplish their work, and that collaboration is important for designing solutions to problems. Students will use the EDP, the same process that professional engineers use in their work, in this lesson. A graphic representation of the EDP is provided at the end of this lesson (p. 84). You may wish to provide each student with a copy of the EDP graphic or enlarge it and post it in a prominent place in your classroom for student reference throughout the module. Be prepared to review each step of the EDP with students and emphasize that the process is not a linear one—at any point in the process, they may need to return to a previous step. The steps of the process are as follows:

1. *Define.* Describe the problem you are trying to solve, identify what materials you can use, and decide how much time and help you have to solve the problem.

2. *Learn.* Brainstorm solutions and conduct research to learn about the problem you are trying to solve.

3. *Plan.* Plan your work, including making sketches and dividing tasks among team members if necessary. Also include safety precautions.

4. *Try.* Build a device, create a system, or complete a product.

5. *Test.* Now, test your solution. This might be done by conducting a performance test, if you have created a device to accomplish a task, or by asking for feedback from others about their solutions to the same problem.

6. *Decide.* Based on what you found out during the Test step, you can adjust your solution or make changes to your device.

After completing all six steps, students can share their solution or device with others. This represents an additional opportunity to receive feedback and make modifications based on that feedback.

The following are additional resources about the EDP:

- *www.sciencebuddies.org/engineering-design-process/engineering-design-compare-scientific-method.shtml*

- *www.pbslearningmedia.org/resource/phy03.sci.engin.design.desprocess/what-is-the-design-process*

COMMON MISCONCEPTIONS

Students will have various types of prior knowledge about the concepts introduced in this lesson. Table 4.6 outlines some common misconceptions students may have concerning these concepts. Because of the breadth of students' experiences, it is not possible to anticipate every misconception that students may bring as they approach this lesson. Incorrect or inaccurate prior understanding of concepts can influence student learning in the future, however, so it is important to be alert to misconceptions such as those presented in the table.

Table 4.6. Common Misconceptions About the Concepts in Lesson 2

Topic	Student Misconception	Explanation
Engineers and the engineering design process (EDP)	All engineers are people who drive or operate trains.	Railroad engineers are just one type of engineer. The engineers referred to in this module are people who use science, technology, and mathematics to build machines, products, and structures that meet people's needs.
	Engineers use only science and mathematics to do their work.	Engineers often use science and mathematics in their work, but they also use other knowledge to solve problems and design products, such as how people use products, what people's needs are, and how the natural environment affects materials.
	Engineers work alone to build things.	Engineers often work in teams and use a process to solve problems. The process involves creative thinking, research, and planning, in addition to building and testing products.

Continued

Table 4.6. (*continued*)

Topic	Student Misconception	Explanation
Sound	As sound waves move through air, the air moves with them, like wind.	Sound waves are vibrating particles that propagate through a medium (i.e., vibrating particles cause neighboring particles to vibrate). They do not cause wind-like movements in air masses.
	Sound can only travel through air.	Sound can travel through a variety of substances.
	Sound cannot travel through solid objects.	Sound can travel through solid objects because the particles of solid objects can vibrate.
	Sound moves more quickly through air than through other substances.	Sound tends to move more quickly through solids and liquids since in these substances particles are more closely packed than in air (i.e., the sound propagates from particle to particle more rapidly). It is important to note, however, that sound does not travel well from one medium to another. For example, the sound made from striking a wall with a hammer will travel quickly through the wall, but many of the sound waves resulting from a loud noise several feet away from the wall will dissipate into the air before reaching the wall.

PREPARATION FOR LESSON 2

Review the Teacher Background Information section (p. 68), assemble the materials for the lesson, duplicate the EDP graphic (p. 84) if you wish to hand it out to students or enlarge it to post in the classroom, and preview the videos recommended in the Learning Components section that follows.

Students will use disposable plastic forks to observe the connection between vibration and sound. The forks should be medium or heavy weight since lightweight plastic will not vibrate as effectively. Test the forks for their vibrating qualities by placing the fork on the edge of a table with the rounded bottom of the fork facing upward. Hold about ¾ of the fork handle firmly on the table with your hand. Draw the thumb of your other hand quickly over the tines of the fork. The fork should visibly vibrate and make a low-pitched sound.

For the mathematics and ELA connections in this lesson, students will consider various ways to describe sounds. For the Introductory Activity/Engagement, prepare a list of sounds in order to assign one to each student team. Choose sounds that students will be familiar with and that represent a range of pitches (e.g., a bird singing, an explosion, a chainsaw, a lion roaring, a woman singing, a dump truck motor). Choose sounds with which students will be familiar since teams will work together to describe these sounds in words.

Have on hand two of the wave graphs you created in Lesson 1. Choose one with a short wavelength and one with a long wavelength to display to students when you introduce the concept of pitch.

Students will use plastic cups with small holes in the bottom for the What's the Buzz? investigation. Prepare for the investigation by making the holes in the bottom center of the cups using a nail or another sharp, pointed instrument. These holes should be large enough for a piece of string to pass through, but small enough so that a knot in the string will not pass through the hole. Prepare an open area for students to use their paper cup telephones. Each student pair will need to be able to stand 5 feet, 10 feet, and 15 feet apart.

Identify a decibel measurement app or software to use for the mathematics connection in this lesson. The following website provides resources for smartphone decibel meter apps: *www.healthyhearing.com/report/47805-The-best-phone-apps-to-measure-noise-levels.*

Students will create estimates in the mathematics connection in this lesson. To help introduce the idea of estimates, students will practice estimating the number of beads or marbles in a container. Prepare *two* plastic containers as follows: Place 10 beads or marbles in one container; count the remaining beads or marbles and fill the second container to the top with these.

American Sign Language (ASL) will be introduced in this lesson. See the National Institute on Deafness and Other Communication Disorders (NIDCD) American Sign Language page at *www.nidcd.nih.gov/health/american-sign-language* for information on ASL and to learn some simple signs.

LEARNING COMPONENTS

Introductory Activity/Engagement

Connection to the Challenge: Begin each day of this lesson by directing students' attention to the module challenge, the Show Me the Waves Challenge:

> *You and your team will be challenged to create a musical show that demonstrates how waves can be used to communicate over distances. You will create musical instruments that create sound waves, and you will also use light to show how sound and light can be used to communicate and entertain.*

Hold a brief class discussion on how students' learning in the previous days' lessons contributed to their ability to complete the challenge. You may wish to create a class list of key ideas on chart paper.

Science Class: Hold a class discussion about sound. Ask students the following questions:

- What are things we can hear?

- How do we hear?

- How does the sound of a voice move from a person's mouth to another person's ears?

Create a KLEWS chart to document student responses and have students use STEM Research Notebook Entry #9 to document their ideas.

STEM Research Notebook Entry #9

Have students document their ideas about sound and hearing in their STEM Research Notebooks, using both words and pictures.

Next, conduct an interactive read-aloud of *What Is Sound?* by Charlotte Guillain. After the read-aloud, ask students to share what they learned about sound, and add students' ideas to the KLEWS chart.

Remind the class that in the last lesson they investigated waves that move through water. Ask students if they are surprised that sound moves in waves and, if so, why this surprises them. Ask students to share their ideas about what evidence we have that sound moves in waves. Next, hand out a plastic fork to each student, instructing students to place the forks on the desk in front of them with the rounded side of the fork facing up. Demonstrate how to make the fork vibrate (see Preparation for Lesson 2, p. 72). Have students duplicate this with their forks. Ask them to share their observations and document these on the KLEWS chart. Next, ask students to experiment by holding less or more of the fork on the table, asking them if the vibrations they see or the sound they hear changes. Record students' observations on the KLEWS chart. Have students complete STEM Research Notebook entry #10 to record their observations. After students have completed their notebook entries, collect the forks.

STEM Research Notebook Entry #10

Have students record their observations about their forks' vibrations and sound in their STEM Research Notebooks, using both words and pictures.

Mathematics and ELA Connections: Ask students to share their ideas about what adjectives they can use to describe sounds, and create a class list of student ideas. Next, have

students assemble in their teams and assign each team a sound (see Preparation for Lesson 2, p. 72). Instruct teams not to tell other teams what their assigned sound is, since later in the lesson other students will guess what the sound was based on the adjectives used to describe it. Have teams brainstorm adjectives to describe their assigned sound, and have each team use STEM Research Notebook Entry #11 to describe their team's assigned sound using words and pictures.

STEM Research Notebook Entry #11

Have students record their descriptions of their teams' assigned sounds in their STEM Research Notebooks, using both words and pictures.

Next, ask students for their ideas about whether sound can be measured. Introduce the term *decibels* as a way that the loudness of a sound can be described using numbers.

Social Studies Connection: Ask students for their ideas about how people use sound to communicate, creating a class list of students' ideas (e.g., speaking, singing, playing musical instruments, radio broadcasts, clapping, knocking on a door, podcasts). Introduce the concept of hearing disabilities to students. Ask them if they know anyone who has a hearing aid or a person who has deafness. Ask students to share their ideas about how people with hearing disabilities communicate with others (e.g., hearing aids to allow them to hear more clearly, cochlear implants, sign language). Create a class list of students' ideas.

Activity/Exploration

Science Class and Mathematics and ELA Connections: Ask students to share their ideas about how their ears help them hear sounds, and record student ideas on a KLEWS chart. Conduct an interactive read-aloud of *The Sense of Hearing,* by Mari Schuh. Then, have students complete STEM Research Notebook Entry #12.

STEM Research Notebook Entry #12

Have students label the parts of the ear using the diagram and word bank provided in STEM Research Notebook Entry #12.

Next, ask students what they do when they want to talk to someone who is in another room (e.g., speak more loudly, yell, move closer to another person). Next, ask students what they can do if they are too far away from someone to hear their voice and cannot move closer to the person (e.g., call them on the telephone). Introduce the idea to students that sound waves can be used to communicate over long distances as well as close up. Conduct an interactive read-aloud of *Sounds All Around,* by Wendy Pfeffer, asking

students to pay particular attention to how sound travels and how sound waves can be used to communicate. After the read-aloud, hold a class discussion about how sound waves can be used to communicate, recording student ideas on a KLEWS chart. Have students document their ideas in STEM Research Notebook Entry #13.

STEM Research Notebook Entry #13

Have students document what they learned about how sound waves can be used to communicate in their STEM Research Notebooks, using both words and pictures.

Have each team share the adjectives they used to describe their assigned sound in the Introductory Activity/Engagement and have the rest of the class make guesses about what their assigned sound was. Record each sound on a class list as it is guessed correctly. After students have guessed each team's sound, ask students to consider what is different about the sounds on the list. Guide students to understand that two sounds can both be loud but can sound different (for example, a woman singing and a lion roaring). Ask students to consider what is different about the sounds (e.g., some are "high" and some are "low"). Introduce the term *pitch* as the highness or lowness of a sound.

Remind students of their wave measurements in Lesson 1. Show students the wave graphs with long and short wavelengths you prepared (see Preparation for Lesson 2, p. 72). Review wavelengths with students. Tell students that these graphs represent how the sound waves vibrate. Remind students of their experiments with the plastic forks in the Introductory Activity/Engagement. Ask students what they observed about how their forks vibrated when they made higher sounds versus lower sounds (e.g., the fork vibrated faster while it made a higher pitched sound). Tell students that the wavelengths of the sound waves tell them about how the particles of a sound waves vibrate. Sound waves with long wavelengths mean that the vibration is slower and the sounds have low pitches, and sound waves with shorter wavelengths mean that the vibration is faster and the sounds have higher pitches.

Ask students to name some low-pitched sounds, creating a class list (e.g., thunder rumbling, a man's voice, a bass drum). Next, ask students to name high-pitched sounds, creating a second list (e.g., a whistle, a small child's voice, shoes squeaking on a tile floor). Next, ask students which of the wave graphs best represents each list (the longer wavelength graph represents the low-pitched sound, the shorter wavelength graph represents the higher-pitched sounds).

Tell students that they will investigate how sound can be used to communicate over distances in the What's the Buzz? investigation.

What's the Buzz? Investigation

Ask students for their ideas about how a telephone works. Guide students to understand that a telephone turns the sound waves they produce with their voices into energy that reaches the person on the other end of the telephone. Introduce the activity by telling students that they are going to create "telephones" that carry the sound waves they produce directly to another person. Show students the materials they will have to create telephones (plastic cups, string, and tape) and hold a class discussion, asking students to share their ideas about how they could use these materials to construct a telephone that they can use to communicate with a partner. Write or sketch student ideas on the board or on chart paper. After students have shared their ideas, tell students that they will use the cups to speak into and to hear through, and that the string will carry the sound waves they create from one cup to the other. Have students connect their ideas with what they know about sound waves and human hearing. Tell students that they will use the POE process for this investigation.

Demonstrate to students how to use the yardstick to measure the string. Have student teams identify the 10-foot piece of string by measuring it. Next, have students pull an end of the string through the holes at the bottom of each cup. Demonstrate to students how to tie a knot in the string so that it does not slip back through the cup (note that students will need to untie these knots later). Instruct each partner to take one cup and walk slowly away from one another until the string is taut. Have one partner hold the cup to his or her ear and have the other partner speak in a whisper into the other cup, then have the partners reverse roles. Have students return to their desks and ask students if they could hear one another.

Ask students the following questions, recording student responses on a POE chart and having students create a STEM Research Notebook entry:

- Will the length of string affect the sound? How? (louder, softer, higher pitch, lower pitch)

- Will the tightness of the string affect the sound? How? (louder, softer, higher pitch, lower pitch)

STEM Research Notebook Entry #14

Have students record their predictions in their STEM Research Notebooks about how the string length and tightness will affect the loudness and pitch of their partner's voice.

Next, have students identify and measure the longer length of string (15 feet) and the shorter length of string (5 feet) and repeat the procedure with each length of string. Instruct students to speak at the same loudness (in a whisper) for each length of string.

For each length of string, students should note whether the sound was louder or softer or the same and whether the pitch was higher or lower or the same and how the tightness of the string changed what they heard.

Record student observations on the POE chart, and have students record observations in STEM Research Notebook Entry #15.

STEM Research Notebook Entry #15

Have students document their observations about the differences in sound for different string lengths and tightness in their STEM Research Notebooks.

Social Studies Connection: Conduct an interactive read-aloud of *A Birthday for Ben*, by Kate Gaynor. Hold a class discussion about difficulties that people with hearing loss may face, asking students the following questions:

- How do you think Ben felt when he went to parties? Why?

- Have you ever felt left out?

- What helped you when you felt that way?

- How did Ben's parents help him?

- What can you do to make sure everyone feels welcome in a group?

Explanation

Science Class: Have students compare their predictions to their observations for the What's the Buzz? investigation and discuss whether or not their predictions were accurate. Ask students to offer explanations for their observations. Record student responses on the POE chart. Students will also document their explanations in their STEM Research Notebooks by completing a cause-and-effect chart. As a class, discuss students' ideas about why their partner's voice was softer with longer string lengths (i.e., the sound waves had further to travel and sound waves lose strength with longer distances) and if they noticed differences in pitch, why this was so (i.e., pitch is determined by the way the string vibrates; shorter strings may vibrate more quickly than longer strings, but this difference may have been too small to be noticeable to students). Likewise, discuss students' explanations for why their partner's voice was softer or impossible to hear with a loose string (i.e., tight strings vibrate more effectively; with the loose string, the vibrations are lost to the air around the string).

STEM Research Notebook Entry #16

Have student record their explanations for the variations in sound they observed in the What's the Buzz? investigation by completing a cause-and-effect chart.

Mathematics, ELA, and Social Studies Connections: Remind students of how they described sounds with words and pictures. Hold a class discussion about how these descriptions could be useful to people who have hearing impairments (e.g., for people who have experience with sounds, the adjectives will describe the sounds) or not useful (e.g., for people who have never had hearing, the adjectives will not be useful).

As a class, discuss the causes of hearing loss (loud noises, physical injury, illness, old age). Focus the class discussion on how loud noises can damage hearing. Emphasize to students that hearing loss can result from sudden loud noises, like explosions or loud sirens close by, and that hearing loss can result from sustained noises over time, like long-term exposure to machinery or loud music. Remind students that loudness is measured in units called decibels and tell students that sounds below 75 decibels are probably safe while sustained exposure to sounds at 85 decibels or above can damage their hearing.

Ask students to name a variety of sounds they think are unsafe for hearing, creating a class list. Next, ask students to name sounds that can be heard in the classroom (e.g., students talking, chair legs scraping on the floor, music being played, school bell, school announcement). Have students copy the list in STEM Research Notebook Entry #17.

Next, introduce the concept of making an estimate to the class, emphasizing that *estimate* means finding a value that is close to the correct answer. Tell students that to make estimates they must have some information about what they are estimating. Show the students the container with 10 beads in it (see Preparation for Lesson 2, p. 72). As a class, count the 10 beads in the container. Now show students the full container of beads. Ask students if they can use the information about what 10 beads in a container look like to make an estimate about how many beads are in the other container. Have students estimate the number of beads in the full container, recording student responses on a class list. After students have made their estimates, tell students the actual number of beads or marbles in the container. Next, create a table with three columns, labeled "Smaller: <," "Same: =," and "Larger: >." Discuss with students what each symbol means. Next, ask students to indicate by a show of hands if they estimated the correct number, if their estimate was too small, or if their estimate was too large. Record the number of students who raise their hands for each. Ask students who estimated the correct number or close to the correct number of beads or marbles to share how they arrived at their estimates.

Tell students that they will now estimate how many decibels of sound each of the classroom sounds on the list creates. Ask students what information they think would be helpful to them to make estimates. Give students the information that a whisper is around 20 decibels and a lawnmower is about 95 decibels. Have students record in STEM Research Notebook Entry #17 an estimate for loudness in decibels of each classroom sound they listed. Next, use the app or decibel measurement software (see Preparation for Lesson 2, p. 72) to measure these sounds and have students record the measurements

in their STEM Research Notebooks. Next, have students record whether their estimates were smaller than (<), the same as (=), or larger than (>) the actual measurements.

STEM Research Notebook Entry #17

Have students record the list of classroom sounds, their estimates for the loudness of the sounds, the actual measurements of the sounds, and the comparison between their estimates and the measurements using <, =, and >.

Ask students for their ideas about how they can protect their hearing when they are around loud noises, creating a class list of student ideas (e.g., avoid loud noises, wear ear protection such as earphones that block noises).

Elaboration/Application of Knowledge

Science Class: Tell students that they will use what they have learned about sound to create musical instruments in the I Can Hear the Music activity. Tell students that in this activity they will act as engineers. Ask students for their ideas about what engineers do, creating a class list of students' ideas. Introduce the idea to students that engineers use the engineering design process (EDP) to do their work. Introduce the steps of the EDP using the EDP graphic provided on page 84.

Next, introduce the idea to students that engineers often work in groups or teams to do their work and that they will work in teams to create their musical instruments and plan their musical show for the Show Me the Waves Challenge. Introduce the term *collaboration*. On the board or on chart paper, create a T-chart titled "collaboration," with one column labeled "benefits" and the other column labeled "challenges." Ask students to share their ideas about why engineers collaborate (e.g., having many people's ideas to draw on, sharing work, being able to talk about ideas, having people with different skills and strengths working on a project). Then, ask students to think about experiences they have had in working with groups and ask them what challenges they faced (e.g., too many people trying to talk at once, one person doing all the work, people not listening to each other). Emphasize to students that collaborating requires all team members to participate and be respectful of each other. Have students work in their teams to brainstorm some rules or guidelines for good collaboration that will help their team work together well. After teams have each identified two or three collaboration rules or guidelines, have each team share their ideas to create a class list. Based on students' ideas, create a class list for good collaboration that incorporates students' ideas into several rules. Post these collaboration rules in the classroom.

I Can Hear the Music Activity

Tell students that their challenge in this activity is to create musical instruments that work by creating vibrations on a surface (essentially a drum) but that also have additional features to make music (for example, making a rattling sound when shaken). Show students the materials each team will have available and display the materials where students can see them as they plan their designs. Each team should make two instruments.

As a class, work through each step of the EDP:

1. *Define.* Ask students to identify the problem they will solve and record student responses on chart paper (create an instrument that creates vibrations on a surface and also has additional interesting musical features).

2. *Learn.* Ask students what they need to know to solve the problem and record student responses on chart paper (how a drum works [the player strikes a material that vibrates] and what kind of materials they have available that would work as a vibrating surface [the balloon, rubber bands]). Then, have students assemble in their teams and brainstorm ideas for their instruments. Distribute one sheet of white paper to each student so that students can sketch designs and record their ideas.

3. *Plan.* Distribute the materials to each team. Tell students that they should not yet start building their instruments, but that they can handle the materials as they decide on their instrument designs. Have student teams discuss their ideas and decide on two instrument designs. After teams have decided on their designs, have each student draw sketches, including labels for what materials they will use, of the two instruments in STEM Research Notebook Entry #18. Remind students that they have specific materials to work with and their sketches should reflect this.

STEM Research Notebook Entry #18

Have each student create labeled sketches for the two instruments their team decided to make.

4. *Try.* Have student teams build their designs.

5. *Test/Decide.* Have teams test their instruments and decide if they work as intended and what, if anything, they can do to improve them.

6. *Share.* Have each team demonstrate their instruments for the class.

Students will use these instruments again in Lesson 4, so you should keep them in the classroom.

Assess student learning for the lesson using the assessment instrument provided in Appendix B (p. 163).

Mathematics, ELA, and Social Studies Connections: Ask students if there are any ways to experience sound without hearing it with their ears, and record student responses on a class list. Next, ask students to lightly place their hands on their throats and make a humming noise. Ask them what they felt (vibrations). Ask students if they would be able to feel those vibrations without hearing themselves hum. Explain to students that they are feeling the vibrations associated with sound waves, and that people who have hearing loss can use vibrations like these to feel music and other sounds. Show a video about a deaf dancer such as the video about Shaheem Sanchez, "How This Deaf Dancer Hears the Music," at *www.youtube.com/watch?v=p0IYSKpADMM*. After viewing the video, hold a class discussion about how this dancer is able to dance with the music without hearing the music with his ears. Next, turn on the radio to play music and have students place their hands over the speaker to see if they can feel the music like the dancer in the video.

Ask students for their ideas about how deaf people communicate with others. Introduce the idea that many people with deafness are able to speak, but that since they cannot hear others' voices, they need a special language to understand what others' are saying to them. Introduce American Sign Language (ASL) as a way to communicate with people who have hearing losses. Teach students some basic word signs (for example, hello, goodbye, friend, teacher, student, and school). You may also wish to teach students some basic phrases (for example, how are you?, what time is it?, and I am tired). Have students work in pairs to practice signing to one another and interpreting the signs.

Next, ask students for their ideas about how numbers are expressed in ASL. Show students a video about how to sign numbers in ASL (for example, "Numbers 1 to 30: ASL" at *www.youtube.com/watch?v=hFCXyB6q2nU*) and have students practice the signs for 1–10. Once students are able to sign these numbers independently, ask them to provide answers to some basic arithmetic facts with sums of 10 or less with the appropriate signs.

Evaluation/Assessment

Students may be assessed on the following performance tasks and other measures listed.

Performance Tasks

- What's the Buzz? communication devices
- I Can Hear the Music instruments
- Lesson 2 Assessment (see Appendix B, p. 163)

Other Measures (see assessment rubric in Appendix B, p. 170)

- Teacher observations

- STEM Research Notebook entries

- Participation in teams during investigations

INTERNET RESOURCES

Pitch and frequency of sound waves
- *www.physicsclassroom.com/class/sound/Lesson-2/Pitch-and-Frequency*

Human ear anatomy and the science of hearing
- *www.physicsclassroom.com/class/sound/Lesson-2/The-Human-Ear*

- *www.asha.org/public/hearing/How-We-Hear*

Types and causes of hearing loss
- *www.nidcd.nih.gov/health/hearing-ear-infections-deafness*

Engineering
- *www.engineergirl.org/33/TryOnACareer*

- *www.nacme.org/types-of-engineering*

- *www.sciencekids.co.nz/sciencefacts/engineering/typesofengineeringjobs.html*

EDP
- *www.sciencebuddies.org/engineering-design-process/engineering-design-compare-scientific-method.shtml*

- *www.pbslearningmedia.org/resource/phy03.sci.engin.design.desprocess/what-is-the-design-process*

Resources for decibel measurement
- *www.healthyhearing.com/report/47805-The-best-phone-apps-to-measure-noise-levels*

ASL information from NIDCD
- *www.nidcd.nih.gov/health/american-sign-language*

"How This Deaf Dancer Hears the Music" video
- *www.youtube.com/watch?v=p0IYSKpADMM*

"Numbers 1 to 30: ASL" video
- *www.youtube.com/watch?v=hFCXyB6q2nU*

ENGINEERING DESIGN PROCESS

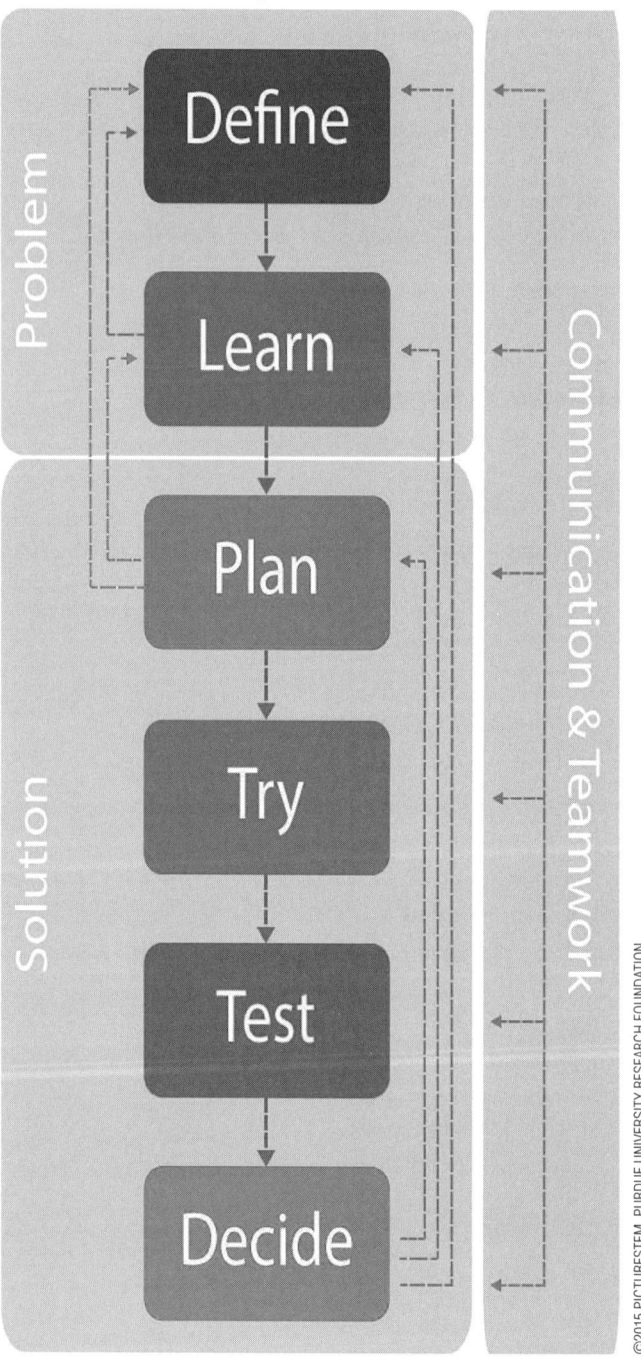

©2015 PICTURESTEM, PURDUE UNIVERSITY RESEARCH FOUNDATION.

Figure 4.3. Anatomy of the Human Ear

ANATOMY OF THE EAR

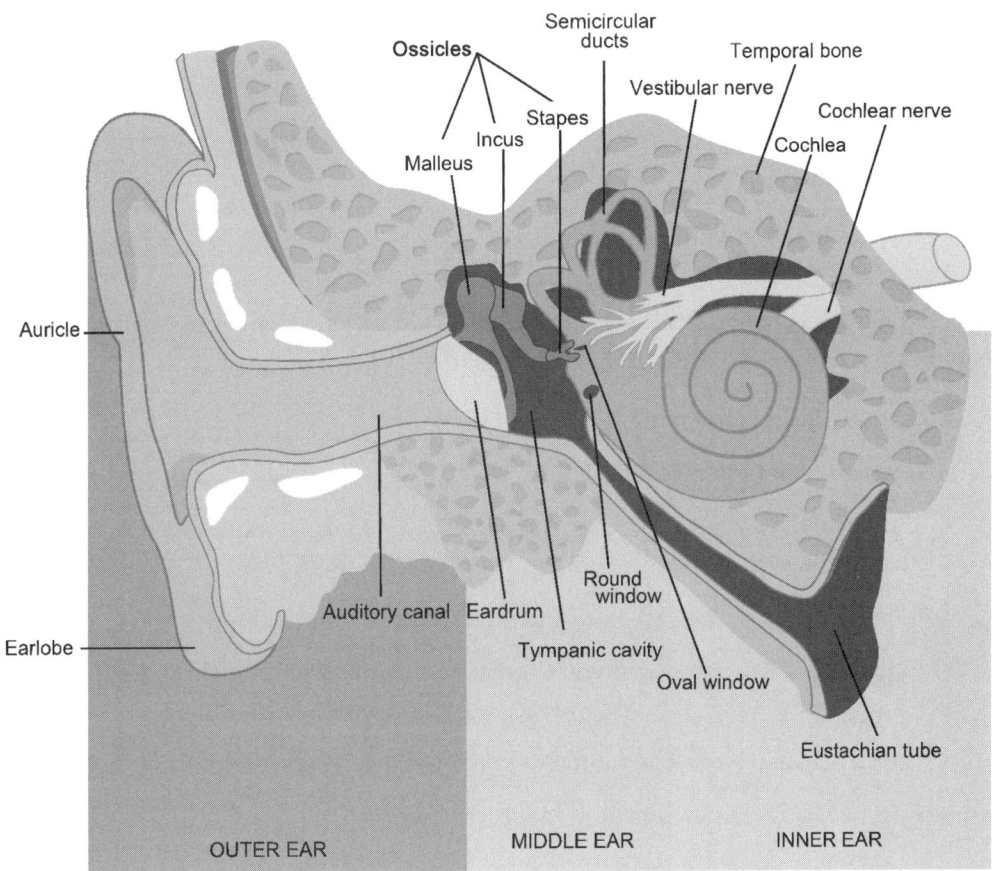

Note: A full-color version of this image is available on the book's Extras page at *www.nsta.org/roadmap-waves.*

Lesson Plan 3: Lighting It Up!

In this lesson, students explore the concept of light through interactive read-alouds, viewing videos, and an inquiry activity in which they investigate how light interacts with various materials.

ESSENTIAL QUESTIONS

- What is light?

- What are things we can see?

- How do we see?

- How can we communicate with light?

ESTABLISHED GOALS AND OBJECTIVES

At the conclusion of this lesson, students will be able to do the following:

- Identify the Sun as a natural source of light

- Describe how light reaches our eyes

- Understand how human eyes respond to light via a basic exploration of the human anatomy of sight

- Identify materials that are transparent, translucent, and opaque

- Explain concepts associated with light when working on models to demonstrate how humans experience and interact with light waves

- Identify several ways that humans experience and interact with light waves

- Describe and demonstrate how shadows are formed

- Explain how light can be used to communicate over distances

- Demonstrate an understanding of the use of descriptive words and imagery in art and literature

TIME REQUIRED

- 8 days (approximately 30 minutes each day; see Tables 3.8–3.10, pp. 38–39)

MATERIALS

Required Materials for Lesson 3

- STEM Research Notebooks

- Computer with internet access for viewing videos

- Books

  - *All About Light*, by Lisa Trumbauer (Children's Press, 2004)

  - *Day Light, Night Light*, by Franklyn M. Branley (HarperCollins, 1998)

  - *Eye: How It Works*, by David Macaulay (Macmillan, 2013)

  - *Light Is All Around Us*, by Wendy Pfeffer (HarperCollins, 2015)

  - *My Shadow*, by Robert Louis Stevenson, illustrated by Sara Sanchez (Sky Pony, 2016; the poem can also be accessed by searching online)

  - *Sending Messages With Light and Sound*, by Jennifer Boothroyd (Lerner, 2014)

  - *Six Dots: A Story of Young Louis Braille*, by Jen Bryant (Knopf Books for Young Readers, 2016)

  - *The Black Book of Colors*, by Menena Cottin (Groundwood Books, 2008)

  - *Thomas Edison and His Bright Idea*, by Patricia Brennan Demuth (Penguin, 2016)

  - *What Makes a Shadow?*, by Clyde Robert Bulla (HarperCollins, 1994)

- Chart paper

- Markers

- Plain white paper (1 sheet per student)

- Crayons (1 set of 8 or more per student)

- Teaching clock or wall clock with moveable hands to demonstrate various times

- Images of the outdoors that reflect local geography and images of the outdoors that reflect nonlocal geography (1 per pair—see Preparation for Lesson 3, p. 93)

- Safety glasses with side shields or safety goggles (per student)

Additional Materials for Lighting the Way (per pair of students)

- Colored lightbulb picture from STEM Research Notebook Entry #21

- 1 piece of clear plastic or plastic wrap, about 6 inches × 6 inches

- 1 piece of wax paper, about 6 inches × 6 inches

- 1 piece of cardboard, about 6 inches × 6 inches

- 1 flashlight

Additional Materials for Shady Shadow Puppets (per team of 4 students)

- 1 large puppet template on 9 inch × 12 inch construction paper

- 3 small puppet templates, on half sheet (4 ½ inch × 6 inch) construction paper

- Glue or tape

- 4 scissors

- 4 large craft sticks or plastic straws

- 2 flashlights

SAFETY NOTES

1. Remind students that safety glasses with side shields or safety goggles must be worn during all phases of this inquiry activity.

2. Immediately pick up any items dropped on the floor to avoid a slip or trip-and-fall hazard.

3. Students should use caution when working with sharps (e.g., scissors, sticks, straws) as they can cut or puncture skin.

4. Students should have direct adult supervision if they are working with scissors.

5. Students should use caution if moving around in the room with light levels lowered because of potential trip-and-fall hazards.

6. Have students wash their hands with soap and water after the activity is completed.

CONTENT STANDARDS AND KEY VOCABULARY

Table 4.7 lists the content standards from the *NGSS, CCSS,* NAEYC, and the Framework for 21st Century Learning that this lesson addresses, and Table 4.8 (p. 92) presents the key vocabulary. Vocabulary terms are provided for both teacher and student use. Teachers may choose to introduce some or all of the terms to students.

Table 4.7. Content Standards Addressed in STEM Road Map Module Lesson 3

NEXT GENERATION SCIENCE STANDARDS

PERFORMANCE EXPECTATIONS

- 1-PS4-2. Make observations to construct an evidence-based account that objects can be seen only when illuminated.

- 1-PS4-3. Plan and conduct an investigation to determine the effect of placing objects made with different materials in the path of a beam of light.

SCIENCE AND ENGINEERING PRACTICES

Planning and Carrying Out Investigations

Planning and carrying out investigations to answer questions or test solutions to problems in K–2 builds on prior experiences and progresses to simple investigations, based on fair tests, which provide data to support explanations or design solutions.
- Plan and conduct investigations collaboratively to produce evidence to answer a question.

Constructing Explanations and Designing Solutions

Constructing explanations and designing solutions in K–2 builds on prior experiences and progresses to the use of evidence and ideas in constructing evidence-based accounts of natural phenomena and designing solutions.
- Make observations (firsthand or from media) to construct an evidence-based account for natural phenomena.

DISCIPLINARY CORE IDEAS

PS4.B: Electromagnetic Radiation

- Objects can be seen if light is available to illuminate them or if they give off their own light.

- Some materials allow light to pass through them, others allow only some light through and others block all the light and create a dark shadow on any surface beyond them, where the light cannot reach. Mirrors can be used to redirect a light beam. (Boundary: The idea that light travels from place to place is developed through experiences with light sources, mirrors, and shadows, but no attempt is made to discuss the speed of light.)

PS4.C: Information Technologies and Instrumentation

- People also use a variety of devices to communicate (send and receive information) over long distances.

CROSSCUTTING CONCEPT

Cause and Effect

- Simple tests can be designed to gather evidence to support or refute student ideas about causes.

Continued

Table 4.7. (*continued*)

COMMON CORE STATE STANDARDS FOR MATHEMATICS

MATHEMATICAL PRACTICES

- MP1. Make sense of problems and persevere in solving them.
- MP2. Reason abstractly and quantitatively.
- MP3. Construct viable arguments and critique the reasoning of others.
- MP4. Model with mathematics.
- MP5. Use appropriate tools strategically.
- MP6. Attend to precision.
- MP7. Look for and make use of structure.
- MP8. Look for and express regularity in repeated reasoning.

MATHEMATICAL CONTENT

- 1.NBT.B.3. Compare two two-digit numbers based on meanings of the tens and ones digits, recording the results of comparisons with the symbols >, =, and <.

COMMON CORE STATE STANDARDS FOR ENGLISH LANGUAGE ARTS

READING STANDARDS

- RI.1.1. Ask and answer questions about key details in a text.
- RI.1.2. Identify the main topic and retell key details of a text.
- RI.1.3. Describe the connection between two individuals, events, ideas, or pieces of information in a text.
- RI.1.7. Use the illustrations and details in a text to describe its key ideas.

WRITING STANDARDS

- W.1.2. Write informative/explanatory texts in which they name a topic, supply some facts about the topic, and provide some sense of closure.
- W.1.6. With guidance and support from adults, use a variety of digital tools to produce and publish writing, including in collaboration with peers.
- W.1.7. Participate in shared research and writing.
- W.1.8. With guidance and support from adults, recall information from experiences or gather information from provided sources to answer a question.

Continued

Table 4.7. (*continued*)

SPEAKING AND LISTENING STANDARDS

- SL.1.1. Participate in collaborative conversations with diverse partners about *grade 1 topics and texts* with peers and adults in small and larger groups.

- SL.1.1.A. Follow agreed-upon rules for discussions.

- SL.1.1.B. Build on others' talk in conversations by responding to the comments of others through multiple exchanges.

- SL.1.1.C. Ask questions to clear up any confusion about the topics and texts under discussion.

- SL.1.3. Ask and answer questions about what a speaker says in order to gather additional information or clarify something that is not understood.

- SL.1.5. Add drawings or other visual displays to descriptions when appropriate to clarify ideas, thoughts, and feelings.

NATIONAL ASSOCIATION FOR THE EDUCATION OF YOUNG CHILDREN STANDARDS

- 2.E.1. Arrange firsthand, meaningful experiences that are intellectually and creatively stimulating, invite exploration and investigation, and engage children's active, sustained involvement by providing a rich variety of material, challenges, and ideas.

- 2.F.3. Extend the range of children's interests and the scope of their thought, present novel experiences and introduce stimulating ideas, problems, experiences, or hypotheses.

- 2.F.6. Enhance children's conceptual understanding through various strategies, including intensive interview and conversation, encourage children to reflect on and "revisit" their experiences.

- 2.G.2. Scaffolding takes on a variety of forms.

- 2.J.1. Incorporate a wide variety of experiences, materials and equipment, and teaching strategies to accommodate the range of children's individual differences in development, skills and abilities, prior experiences, needs, and interests.

- 3.A.1. Teachers consider what children should know, understand, and be able to do across the domains.

FRAMEWORK FOR 21ST CENTURY LEARNING

- Interdisciplinary Themes; Learning and Innovation Skills; Information, Media, and Technology Skills; Life and Career Skills

---

Table 4.8. Key Vocabulary in Lesson 3

Key Vocabulary	Definition
absorb	to take in or soak up
blindness	a condition in which people can see no light or very little light
braille	a type of writing for people with blindness or visual impairment that uses raised dots to represent letters
imagery	describing language that calls on the listener to use his or her senses to create a mental picture
light	a form of energy that reflects from objects into our eyes to let us see
opaque	adjective that describes an object that does not allow light shining on it to pass through
reflection	a wave's change in direction of travel when it encounters a change in medium
shadow	a dark shape formed when light is blocked
translucent	adjective that describes an object that allows some light shining on it to pass through
transparent	adjective that describes an object that allows light shining on it to pass through
visual impairment	a condition in which a person's sight is negatively affected

TEACHER BACKGROUND INFORMATION

Students continue their investigation of waves in this lesson by exploring light, another wave phenomenon.

The Electromagnetic Spectrum and Human Vision

Unlike the waves investigated in previous lessons, light waves are a type of electromagnetic wave. These electromagnetic waves can travel through space (a vacuum) unlike the mechanical waves investigated in Lessons 1 and 2. Microwaves, radio waves, and x-rays are examples of other types of electromagnetic waves. While this lesson makes the connection that light, like sound, travels in waves, the emphasis in this lesson is on the behavior of visible light as it interacts with substances and the human eye. For more information about light waves and electromagnetic waves, see NASA's video tour of the electromagnetic spectrum at *https://science.nasa.gov/ems/09_visiblelight*.

The anatomy of vision is a complex process that requires an understanding of anatomy that is beyond the scope of a first-grade lesson; however, you may introduce basic eye anatomy as part of this lesson. Kids Health offers diagrams and a narrated article at *www.kidshealth.org/en/kids/eyes.html#* that may be useful.

You may wish to make a connection to braille as a resource for people who are blind. The American Printing House for the Blind provides information about braille at *https://braillebug.org*.

COMMON MISCONCEPTIONS

Students will have various types of prior knowledge about the concepts introduced in this lesson. Table 4.9 outlines some common misconceptions students may have concerning these concepts. Because of the breadth of students' experiences, it is not possible to anticipate every misconception that students may bring as they approach this lesson. Incorrect or inaccurate prior understanding of concepts can influence student learning in the future, however, so it is important to be alert to misconceptions such as those presented in the table.

Table 4.9. Common Misconceptions About the Concepts in Lesson 3

Topic	Student Misconception	Explanation
Light	Light does not exist independently in space; it does not travel, but rather comes from a source and affects objects at the same time.	Light exists independently and travels through space.
	Only shiny objects reflect light.	All objects both reflect and absorb light in varying degrees.
	The Moon is a source of light.	The Moon shines because it reflects the Sun's light from its surface.
	Shadows are reflections from the Sun.	Shadows occur when an object comes between light and a surface, blocking the light from reaching the surface.

PREPARATION FOR LESSON 3

Review the Teacher Background Information section, assemble the materials for the lesson, and preview the video recommended in the Learning Components section that follows.

You will introduce the Shady Shadow Puppets activity (see Activity/Exploration) by taking students outside on a sunny day to observe shadows and play shadow tag. Check the weather report and prepare accordingly. For the Shady Shadow Puppets activity, be prepared to group students in teams of four and prepare a space that is large enough for each team to stand 10 feet from a blank wall with no obstacles between the students and the wall. This space should be dark enough so that students can see their puppets' shadows on the wall when they shine a flashlight on the puppet. For each team, use masking tape to mark a spot 10 feet from the wall and a spot 3 feet from the wall. You should also copy puppet templates onto construction paper (templates for large and small puppets are provided at the end of this lesson on pp. 105–106).

In the Explanation section of this lesson, students will work in pairs to describe outdoor images to one another and draw their own versions of these pictures based on these descriptions. Each student pair will need to have one image that reflects local geography and one that reflects geography that is not common in your local area. Identify and print or copy the two images for each student pair. You may use the same two images for each student pair or you may wish to use different sets of images for different student pairs.

LEARNING COMPONENTS
Introductory Activity/Engagement

Connection to the Challenge: Begin each day of this lesson by directing students' attention to the module challenge, the Show Me the Waves Challenge:

> *You and your team will be challenged to create a musical show that demonstrates how waves can be used to communicate over distances. You will create musical instruments that create sound waves, and you will also use light to show how sound and light can be used to communicate and entertain.*

Hold a brief class discussion on how students' learning in the previous days' lessons contributed to their ability to complete the challenge. You may wish to create a class list of key ideas on chart paper.

Science Class and ELA Connection: Remind students that part of their challenge for this module is to incorporate light into their musical shows. Hold a class discussion about light, asking students the following:

- What is light?

- How does light help us see?

- What are things we can see?

- How do we see?

- Where does light come from?

Document student responses on a KLEWS chart. Students will also record their ideas in STEM Research Notebook Entry #19.

STEM Research Notebook Entry #19

Have students document their ideas about light in their STEM Research Notebooks, using both words and pictures.

Next, conduct an interactive read-aloud of *All About Light*, by Lisa Trumbauer. After the reading, discuss the questions above again, adding students' ideas to the KLEWS chart. Students will also use STEM Research Notebook Entry #20 to document what they learned about light.

STEM Research Notebook Entry #20

Have students record what they learned about light in their STEM Research Notebooks, using both words and pictures.

Tell students that light, like sound, also travels in waves, but that light waves do not cause vibrations that we can easily feel, and we cannot see light waves. Instead, we experience light through our vision and through our skin absorbing light and making us warm.

Mathematics Connection: Turn off the lights in the classroom and then turn on a flashlight. Ask students how long it took for the light to travel from the flashlight from their eyes. Introduce the idea that light travels from a source, like the Sun or a lightbulb, to our eyes, but that it travels so quickly that it seems like it takes no time at all. Ask students if they think there is anything in the world that travels faster than light. Next, conduct an interactive read-aloud of *Day Light, Night Light*, by Franklyn M. Branley, asking students to pay close attention to information about how light travels.

Social Studies Connection: Turn off the classroom lights and ask students whether things in the classroom look different or the same. Next, have students cover their eyes with their hands to block out the light. Ask students what they see. Tell students that light enters their eyes to help them see. Ask students to share ideas about how people use sight on a daily basis. Introduce the concept of sight disabilities to students. Ask them if they know anyone who wears glasses or a person who has blindness. Ask students to share their ideas about how people with sight disabilities compensate for this (e.g., glasses, braille books).

Activity/Exploration

Science Class and Mathematics Connection: Students will investigate light through two activities in this lesson, Lighting the Way and Shady Shadow Puppets.

Lighting the Way

Introduce the Lighting the Way investigation by turning off the lights in the classroom and shining a flashlight. Ask students if there is anything they can do to keep the light from getting to their eyes (closing their eyes, covering the flashlight with something). Have students close their eyes. Ask them if they can still see any light from the flashlight. First, introduce to students that they will use flashlights for this activity. Demonstrate to students how to use the flashlights and instruct them not to shine the light in their own or other students' eyes. Ask students for their ideas about where the light comes from in a flashlight (a lightbulb powered by batteries). Ask students what they know about lightbulbs, asking the following questions:

- How do we use lightbulbs in our everyday lives?

- Who invented lightbulbs?

- How did people light their homes before there were lightbulbs?

Record student ideas on a KLEWS chart. Next, conduct an interactive read-aloud of *Thomas Edison and His Bright Idea,* by Patricia Brennan Demuth. After the read-aloud, hold a class discussion about what students learned about lightbulbs and Thomas Edison, recording student responses on the KLEWS chart. Have students complete STEM Research Notebook Entry #21, in which they will color a lightbulb picture to use in the Lighting the Way activity.

STEM Research Notebook Entry #21

Have students color a lightbulb picture and identify the inventor of the lightbulb in their STEM Research Notebooks.

Next, ask students if light can pass through solid objects and have students give examples of solids that light can pass through (e.g., glass) and solids that light does not pass through (e.g., a wall).

Tell students that in the Lighting the Way investigation they are going to explore how light passes through (or doesn't pass through) various materials. Students will use the POE process in the investigation. Have students work in pairs for this activity in which they will experiment with shining light through clear plastic, wax paper, and cardboard placed over their lightbulb pictures from STEM Research Notebook Entry #21.

Show students the materials they will work with. Have students make predictions about how light will interact with various materials, and record student ideas on a POE chart, asking the following questions:

- If you place clear plastic over your lightbulb picture, will you be able to see the picture when you shine a flashlight on it?

- If you place wax paper over your lightbulb picture, will you be able to see the picture when you shine a flashlight on it?

- If you place cardboard over your lightbulb picture, will you be able to see the picture when you shine a flashlight on it?

Document student predictions on the POE chart and have students document their predictions in STEM Research Notebook Entry #22.

STEM Research Notebook Entry #22

Have students record their predictions about how light will pass through the three materials.

Next, have students conduct their observations. Darken the room and direct the student pairs to place the clear plastic over one partner's lightbulb picture and record their observations. Next, have students place the waxed paper over one partner's lightbulb picture and record their observations. Finally, have students place the cardboard over one partner's lightbulb picture and record their observations. The students should record all of these observations in STEM Research Notebook Entry #23.

STEM Research Notebook Entry #23

Have students document their observations about how light passes through each of the three materials.

Discuss student observations as a class and add to the POE chart by asking students the following questions:

- Did you see your picture through clear plastic when you shined a flashlight on it?

- Did you see your picture through wax paper when you shined a flashlight on it?

- Did you see your picture through cardboard when you shined a flashlight on it?

Next, introduce the terms *transparent, translucent,* and *opaque.* Have students identify which term describes the clear plastic (transparent), the wax paper (translucent), and the cardboard (opaque). Then, create a class chart with three columns labeled with each of

the terms and have students identify other materials for each, adding student ideas to the chart.

As a class, revisit predictions from before the Lighting the Way investigation. Have students compare their predictions with their observations. Ask students to explain why they think that they could or could not see the picture when they shined a flashlight through each of the three materials. Discuss whether their predictions were accurate, close, or not accurate and why, and document student explanations on the POE chart. Next, have students complete STEM Research Notebook Entry #24.

STEM Research Notebook Entry #24

Have students record their explanations for why light shone differently through the materials, using the terms *transparent, translucent,* and *opaque.*

Shady Shadow Puppets

Introduce the Shady Shadow Puppets activity by taking students outdoors on a sunny day. Ask them to observe their shadows. Have students change positions and ask them if and how their shadows change. Ask students to offer ideas about why they have a shadow. Have students play a game of shadow tag in which the goal is to tag their classmates' shadows (with no physical touching).

Return to the classroom and hold a class discussion about what students observed about shadows, recording student ideas on a KLEWS chart. Next, conduct an interactive read-aloud of *What Makes a Shadow?*, by Clyde Robert Bulla. After the read-aloud, hold a class discussion about what students learned about shadows, adding to the KLEWS chart.

Tell students that in the Shady Shadow Puppets activity they will work in groups of four to create puppets for a puppet show, but that the audience will watch the puppets' shadows. Ask students how puppets for a shadow puppet show might be different than puppets in a show where the audience sees the actual puppets (e.g., in a shadow puppet show the audience will see only the puppets' outline so the puppets don't need to be colored). Students will use what they have learned about puppets and the POE process to create their puppets and explore their puppets' shadows.

Ask students to predict how their puppets' shadows will behave based on their observations and the read-aloud, asking students the following questions:

- Will the size of a puppet affect the shadow size?

- Will the distance the puppet is held from the wall affect the shadow size?

- Will the distance the puppet is from the light source (a flashlight) affect the shadow size?

Document student responses on a POE chart. Students will also record their predictions in STEM Research Notebook Entry #25.

STEM Research Notebook Entry #25

Have students record their predictions about how the size of their puppets and the distance from the wall and from the light source will affect their puppets' shadows.

Group students in teams of four. Have each student on each team draw and cut the puppets out of construction paper, taping a craft stick or straw to the back of the puppets as handles. Instruct students to choose one team member to create a large puppet (using the full-sized sheet of construction paper for this puppet) while the other team members create smaller puppets (using the half-sized sheets of construction paper). After students have created their puppets, do the following:

- Have the students on each team compare their puppets to make sure that one is larger than the others.

- Have students stand 10 feet from the wall. One student from each team should hold the flashlight at this spot pointed toward the wall and a second student should hold the largest puppet a few inches from the flashlight's beam so the puppet's shadow is cast on the wall. The students should observe whether the shadow is larger than, smaller than, or the same size as the puppet and record their observations in STEM Research Notebook Entry #26.

- Have a third student from each team hold the second flashlight 10 feet from the wall and have the fourth member of each team hold one of the smaller puppets a few inches from the flashlight's beam so the puppet's shadow is cast on the wall. Students should compare the sizes of the shadows of the two puppets and record their observations in STEM Research Notebook Entry #26.

- Next, have the students holding the flashlights remain at 10 feet from the wall, and have the students holding both the large and small puppets slowly move the puppets in the flashlight beam from near the flashlight to the point that is 3 feet from the wall. Students should record their observations in STEM Research Notebook Entry #26.

STEM Research Notebook Entry #26

Have students document their observations about their puppets' shadows in their STEM Research Notebooks.

After students have created their puppets and collected their data, hold a class discussion about students' observations, recording student responses on a POE chart. Ask the following questions during the discussion:

- Did the size of the puppet affect the shadow size?

- Did the distance the puppet was held from the wall affect the shadow size?

- Did the shadow get bigger or smaller when the puppet was closer to the wall and further away from the light source?

As a class, revisit the predictions from before the Shady Shadow Puppets investigation. Discuss whether students' predictions were accurate, close, or not accurate and why. Have students offer explanations for their observations and record them on the POE chart.

ELA Connection: Discuss with students what elements are usually included in a puppet show (e.g., there is usually a story being told or information being given, characters, and dialogue). Have student teams work together to plan a 2-minute shadow puppet show that teaches other students what they have learned about shadows. Have each team share its shadow puppet show with the class.

Social Studies Connection: Hold a class discussion about resources available for people who are blind. Introduce braille as a code that helps blind people read books in many languages, solve math problems, and read music. Conduct an interactive read-aloud of *Six Dots: A Story of Young Louis Braille,* by Jen Bryant.

Explanation

Science Class: Introduce the basic anatomy of the eye to students with a focus on the concept that vision is a complex process that requires many different parts of the eye to interact with incoming light. Ask students what they know about how people see, recording student ideas on a KLEWS chart. Next, conduct an interactive read-aloud of *Eye: How It Works,* by David Macaulay. After the read-aloud, ask students to share what they learned. Record student responses on a KLEWS chart, and have students complete STEM Research Notebook Entry #27.

STEM Research Notebook Entry #27

Have students record what they learned about eye anatomy in their STEM Research Notebooks.

As students complete the entry, you may wish to project a basic eye anatomy image from Macaulay's book or a resource such as *http://kidshealth.org/en/kids/eyes.html#.*

Next, conduct an interactive read-aloud of *Light Is All Around Us,* by Wendy Pfeffer, asking students to pay attention to how they see light interacting with people's eyes. Discuss what students learned about light and how they might be able to use what they learned as they create their musical shows for the Show Me the Waves Challenge.

STEM Research Notebook Entry #28

Have students document what they learned about light in their STEM Research Notebooks and how they might use this in the Show Me the Waves Challenge, using both words and pictures.

Mathematics and ELA Connections: Conduct an interactive read-aloud of *My Shadow,* by Robert Louis Stevenson (this can be accessed via an Internet search or you can buy the version of book illustrated by Sara Sanchez). Hold a class discussion about when the author saw his shadow in the poem and when it was missing. Ask students to use what they have learned about shadows to explain why the author sometimes saw his shadow and sometimes didn't. Next, tell students that the author wrote this poem without using illustrations. If you used the Sanchez-illustrated book, ask students for their ideas about how the illustrator came up with pictures to use with the poem. Introduce the term *imagery* and tell students that as they read text without pictures that they can create images in their minds, or use their imaginations to create a picture of what is happening; this is how illustrators work to illustrate a story.

Tell students that they are going to use what they know about shadows to create images in their minds that they will then draw on paper. Give each student a piece of plain white paper and have them fold the paper in half along the short side of the paper and create a crease. Have students unfold the paper. Tell students to imagine that they are at the beach at 10 a.m. on a sunny summer day and are looking for their shadows and the cause of their shadows (i.e., the sun). Tell students to create a picture in their minds about how this might look. Have students draw and label a picture of themselves in the top section of the paper that depicts the image they created in their minds about how their shadow and the Sun would look. After students have completed their pictures, tell them to now imagine that they are on the same beach at 8 p.m. and that the Sun has just gone down, but there is still enough light to see. Have students draw and label a picture of themselves in the bottom section of the paper that depicts the image they created in their minds about how their shadow would look.

Hold a class discussion about what time of day students can see their shadows and when they cannot. Connect student ideas to time, showing students various times on a clock (e.g., 10 a.m., 10 p.m.) and asking them if they could see their shadow. Then ask students to name when their shadows will be longest (sunrise and sunset). As a class,

research what time sunrise and sunset is at this time of year and have students work as a class to demonstrate these times on a clock.

Social Studies and ELA Connections: Conduct an interactive read-aloud of *The Black Book of Colors,* by Menena Cottin. This book incorporates pictures with raised lines and braille letters, so you should be prepared to allow students to take turns touching the pages of the book. After the read-aloud, hold a class discussion about how reading this book was different from and the same as reading a full-color book, recording students' responses on a class chart. Remind students about their discussion of imagery from the mathematics and ELA connection above. Ask students how they think that people with blindness and those who are speaking to them can use adjectives and imagery to help explain things we understand by seeing them (e.g., by describing a scene using adjectives, we can help a person with blindness create an image in his or her mind).

Next, group students in pairs. Provide one student in each pair with a picture of an outdoor scene that is typical of the geography where you live and the other student a picture of an outdoor scene that is unlike the geography where you live (see Preparation for Lesson 3, p. 93). Instruct students to leave the pictures face down and not to let their partner see the image. Then, have one student from each pair look at the picture he or she was given (the students should do this without showing their partners) and describe it in words to their partners. The partners listening to the descriptions should use STEM Research Notebook Entry #29 to draw a picture of the image they form. Then, partners should switch roles and repeat the exercise with the other picture. After both partners have had a chance to draw a picture, have students compare their drawings with the actual pictures and discuss how the drawings are like and unlike the actual pictures and why.

STEM Research Notebook Entry #29

Have each student draw the picture that his or her partner describes.

Next, have partners share with the class why it was difficult or easy to draw the picture based on their partners' description. Hold a class discussion about how this information might be useful to them when they are speaking with a person who has sight loss, and create a class list of ideas.

Elaboration/Application of Knowledge

Science Class and ELA Connection: Continue exploring using light and sound to communicate through an interactive read-aloud of *Sending Messages With Light and Sound,* by Jennifer Boothroyd. After the reading, ask students to share their ideas about how sound is used to communicate and how light is used to communicate, creating a class chart to record students' ideas. Ask students to review their responses in STEM Research Notebook Entry #28 about how they could use light in the Show Me the Waves Challenge

musical performance. Ask students if they got any new ideas after the reading and have them add these ideas to their STEM Research Notebook entries.

Have students complete the lesson assessment (see Appendix B, p. 165).

Mathematics Connection: Ask students if all light looks the same to them (for example, does the light of a flashlight look the same as the light on a bright sunny day). Hold a class discussion about students' ideas for this difference. Write a list of types of things that give out light on the board (e.g., the sun, a nightlight, a camera flash, the classroom lights, a candle flame, a flashlight beam, a laser pointer beam, stars at night). As a class, work to order this list from brightest to least bright. Then, use <, =, or > signs to compare items with the list (e.g., a flashlight's brightness is > the sun's brightness; a candle flame's brightness is < the classroom light's brightness).

Social Studies Connection: Hold a class discussion about the achievements of people with vision loss and blindness. For example, show a video about Trevor Thomas, a blind hiker (one such video can be found at *www.youtube.com/watch?v=e-K2B6NhQvg*) or a video about Wanda Diaz-Merced, a blind astrophysicist (a video can be found at *www.nationalgeographic.com/video/shorts/1049215555588*). Hold a class discussion about how blindness affected each person, the obstacles these people faced, how they overcame these obstacles, and what resources they used to help them (e.g., guide dogs, braille books, technology that converts light images into sounds).

Evaluation/Assessment

Students may be assessed on the following performance tasks and other measures listed

Performance Tasks

- Lighting the Way investigation

- Shady Shadow Puppets investigation and show

- Shadow image pictures

- Partner description pictures

- Lesson 3 Assessment (see Appendix B, p. 165)

Other Measures (see assessment rubric in Appendix B, p. 170)

- Teacher observations

- STEM Research Notebook entries

- Participation in teams during investigations

INTERNET RESOURCES

NASA video tour of the electromagnetic spectrum
- *https://science.nasa.gov/ems/09_visiblelight*

Human eye anatomy
- *www.kidshealth.org/en/kids/eyes.html#*

Information about braille from the American Printing House for the Blind
- *https://braillebug.org*

Video about Trevor Thomas
- *www.youtube.com/watch?v=e-K2B6NhQvg*

Video about Wanda Diaz-Merced
- *www.nationalgeographic.com/video/shorts/104921555558*

LARGE PUPPET TEMPLATE

SMALL PUPPET TEMPLATE

Lesson Plan 4: Show Me the Waves Challenge

In this lesson, students synthesize their learning from the previous lessons to address the module challenge. Students demonstrate their learning about light and sound waves by creating a musical performance that incorporates light. Students also use the EDP to create additional musical instruments for their performance and explore ways to incorporate light in their performances

ESSENTIAL QUESTIONS

- How can we use our learning about sound waves to create a musical performance?

- How can light be used in a musical show?

ESTABLISHED GOALS AND OBJECTIVES

At the conclusion of this lesson, students will be able to do the following:

- Explain that the pitch of a sound is influenced by properties of the material through which the sound waves move

- Demonstrate their understanding of the EDP by applying it to create musical instruments

- Identify several ways that humans experience and interact with sound and light

- Explain that sound can be used to communicate over distance and create models that demonstrate this

- Explain that light can be used to communicate over distances and create models that demonstrates this

- Demonstrate how humans can experience and interact with sound and light through a musical performance

TIME REQUIRED

- 4 days (approximately 30 minutes each day; see Table 3.10, p. 39)

MATERIALS

Required Materials for Lesson 4

- STEM Research Notebooks

- Computer with internet access for viewing videos

- Books

  - *A Picture Book of Helen Keller,* by David A. Adler (Holiday House, 1990)

  - *I Know a Shy Fellow Who Swallowed a Cello,* by Barbara S. Garriel (Boyds Mill Press, 2012)

  - *Stand in My Shoes: Kids Learning About Empathy,* by Bob Sornson (Love and Logic, 2013)

- Chart paper

- Markers (1 set of 8 or more per student team of 3–4 students)

- Stopwatches or timers (1 per student team of 3–4 students)

- Poster board (1 piece per student team of 3–4 students)

- Safety goggles or safety glasses with side shields (per student)

Additional Materials for Singing Strings Activity (1 per pair of students)

- 1 empty tissue box—preferably rectangular (not square)

- 1 empty paper towel roll

- 4 nonlatex rubber bands that are different sizes—one #31 (2 ½ inches × ⅛ of an inch), one #32 (3 inches × ⅛ of an inch, one #63 (3 inches × ¼ of an inch), and one #73 (3 inches × ⅜ of an inch)

- Masking tape

- Glue

- Scissors

Additional Materials for Incredible Instruments Activity (1 per team of student)

- Glue

- Masking tape

- Scissors

- Nonlatex rubber bands (variety of widths and lengths)

- Craft materials (construction paper, markers, and other items to decorate musical instruments)

- 1 cup of dry beans or small beads

- Various recyclable materials such as the following:

- Toilet paper and paper towel rolls
- Empty tissue boxes
- Empty plastic containers with lids
- Milk jugs
- Empty cans (ensure there are no sharp edges)
- Wax paper

Lighting Materials for Show Me the Waves Challenge (1 per team)

- 2 flashlights
- Assortment of colored tissue paper, translucent fabric, and colored cellophane
- 2 nonlatex rubber bands (appropriately sized to secure materials over the flashlight lens)
- Scissors

SAFETY NOTES

1. Remind students that safety glasses with side shields or safety goggles must be worn during all phases of this inquiry activity.

2. Immediately pick up any items dropped on the floor (e.g., beans) to avoid a slip or trip-and-fall hazard.

3. Students should use caution when working with sharps (e.g., scissors) as they can cut or puncture skin.

4. Students should have direct adult supervision if they are working with scissors.

5. Students should use caution if moving around in the room with light levels lowered because of potential trip-and-fall hazards.

6. Remind students to not eat any food used in the activity.

7. Have students wash their hands with soap and water after the activity is completed.

CONTENT STANDARDS AND KEY VOCABULARY

Table 4.10 lists the content standards from the *NGSS, CCSS,* NAEYC, and the Framework for 21st Century Learning that this lesson addresses, and Table 4.11 (p. 113) presents the key vocabulary. Vocabulary terms are provided for both teacher and student use. Teachers may choose to introduce some or all of the terms to students.

Table 4.10. Content Standards Addressed in STEM Road Map Module Lesson 4

NEXT GENERATION SCIENCE STANDARDS

PERFORMANCE EXPECTATIONS

- 1-PS4-1. Plan and conduct investigations to provide evidence that vibrating materials can make sound and that sound can make materials vibrate.

- 1-PS4-4. Use tools and materials to design and build a device that uses light or sound to solve the problem of communicating over a distance.

SCIENCE AND ENGINEERING PRACTICES

Planning and Carrying Out Investigations

Planning and carrying out investigations to answer questions or test solutions to problems in K–2 builds on prior experiences and progresses to simple investigations, based on fair tests, which provide data to support explanations or design solutions.
- Plan and conduct investigations collaboratively to produce evidence to answer a question.

Constructing Explanations and Designing Solutions

Constructing explanations and designing solutions in K–2 builds on prior experiences and progresses to the use of evidence and ideas in constructing evidence-based accounts of natural phenomena and designing solutions.
- Make observations (firsthand or from media) to construct an evidence-based account for natural phenomena.

Continued

Table 4.10. (*continued*)

DISCIPLINARY CORE IDEAS

PS4.A: Wave Properties
- Sound can make matter vibrate, and vibrating matter can make sound.

PS4.B: Electromagnetic Radiation
- Objects can be seen if light is available to illuminate them or if they give off their own light.
- Some materials allow light to pass through them, others allow only some light through and others block all the light and create a dark shadow on any surface beyond them, where the light cannot reach. Mirrors can be used to redirect a light beam. (Boundary: The idea that light travels from place to place is developed through experiences with light sources, mirrors, and shadows, but no attempt is made to discuss the speed of light.)

PS4.C: Information Technologies and Instrumentation
- People also use a variety of devices to communicate (send and receive information) over long distances.

CROSSCUTTING CONCEPT

Cause and Effect
- Simple tests can be designed to gather evidence to support or refute student ideas about causes.

COMMON CORE STATE STANDARDS FOR MATHEMATICS

MATHEMATICAL PRACTICES
- MP1. Make sense of problems and persevere in solving them.
- MP2. Reason abstractly and quantitatively.
- MP3. Construct viable arguments and critique the reasoning of others.
- MP4. Model with mathematics.
- MP5. Use appropriate tools strategically.
- MP6. Attend to precision.
- MP7. Look for and make use of structure.
- MP8. Look for and express regularity in repeated reasoning.

MATHEMATICAL CONTENT
- 1.NBT.B.3. Compare two two-digit numbers based on meanings of the tens and ones digits, recording the results of comparisons with the symbols >, =, and <.
- 1.OA.A.2. Solve word problems that call for addition of three whole numbers whose sum is less than or equal to 20, e.g., by using objects, drawings, and equations with a symbol for the unknown number to represent the problem.

Continued

Table 4.10. (*continued*)

COMMON CORE STATE STANDARDS FOR ENGLISH LANGUAGE ARTS

READING STANDARDS

- RI.1.1. Ask and answer questions about key details in a text.
- RI.1.2. Identify the main topic and retell key details of a text.
- RI.1.3. Describe the connection between two individuals, events, ideas, or pieces of information in a text.
- RI.1.7. Use the illustrations and details in a text to describe its key ideas.

WRITING STANDARDS

- W.1.2. Write informative/explanatory texts in which they name a topic, supply some facts about the topic, and provide some sense of closure.
- W.1.6. With guidance and support from adults, use a variety of digital tools to produce and publish writing, including in collaboration with peers.
- W.1.7. Participate in shared research and writing.
- W.1.8. With guidance and support from adults, recall information from experiences or gather information from provided sources to answer a question.

SPEAKING AND LISTENING STANDARDS

- SL.1.1. Participate in collaborative conversations with diverse partners about *grade 1 topics and texts* with peers and adults in small and larger groups.
- SL.1.1.A. Follow agreed-upon rules for discussions.
- SL.1.1.B. Build on others' talk in conversations by responding to the comments of others through multiple exchanges.
- SL.1.1.C. Ask questions to clear up any confusion about the topics and texts under discussion.
- SL.1.3. Ask and answer questions about what a speaker says in order to gather additional information or clarify something that is not understood.
- SL.1.5. Add drawings or other visual displays to descriptions when appropriate to clarify ideas, thoughts, and feelings.

NATIONAL ASSOCIATION FOR THE EDUCATION OF YOUNG CHILDREN STANDARDS

- 2.E.1. Arrange firsthand, meaningful experiences that are intellectually and creatively stimulating, invite exploration and investigation, and engage children's active, sustained involvement by providing a rich variety of material, challenges, and ideas.
- 2.F.3. Extend the range of children's interests and the scope of their thought, present novel experiences and introduce stimulating ideas, problems, experiences, or hypotheses.

Continued

Table 4.10. (*continued*)

- 2.F.6. Enhance children's conceptual understanding through various strategies, including intensive interview and conversation, encourage children to reflect on and "revisit" their experiences.
- 2.G.2. Scaffolding takes on a variety of forms.
- 2.J.1. Incorporate a wide variety of experiences, materials and equipment, and teaching strategies to accommodate the range of children's individual differences in development, skills and abilities, prior experiences, needs, and interests.
- 3.A.1. Teachers consider what children should know, understand, and be able to do across the domains.

FRAMEWORK FOR 21ST CENTURY LEARNING
- Interdisciplinary Themes; Learning and Innovation Skills; Information, Media, and Technology Skills; Life and Career Skills

Table 4.11. Key Vocabulary for Lesson 4

Key Vocabulary	Definition
empathy	the ability to understand and care about another person's feelings
percussion instruments	musical instruments that are played by striking, scraping, or rubbing a surface with a hand or another object
string instruments	musical instruments that are played by moving tightly stretched strings so that they vibrate; the strings can be moved by plucking or striking them or by moving a bow across the strings

TEACHER BACKGROUND INFORMATION

In this culminating lesson, students will synthesize their learning from previous lessons to create musical instruments and plan and create a musical show.

COMMON MISCONCEPTIONS

In this lesson, students will apply their learning about concepts introduced in Lessons 1–3. Review the common misconceptions introduced in those lessons (Tables 4.3, 4.6, and 4.9) and be alert to continued misunderstanding of these concepts.

PREPARATION FOR LESSON 4

Assemble the materials for the lesson and duplicate the student handouts.

Decide whether students will choose songs to perform from a list of music you provide or whether students will write their own songs. If you provide songs for students to perform, you may wish to choose songs that students have learned in music class or songs that are in the public domain such as the alphabet song or classic nursery rhymes, or original classroom songs that you have created.

In the Singing Strings activity, student teams will create tissue box guitars. You should begin collecting tissue boxes and paper towel rolls in advance. Prepare a sample instrument to display to the class, using nonlatex rubber bands of the same width (see Singing Strings activity on p. 116).

In the Incredible Instruments activity, students will use the EDP and recycled materials to create innovative musical instruments. You should begin collecting recyclables in advance or have students bring recyclable items from home. These items may include those noted in the materials list as well as any other recycled containers that are safe for students to work with (i.e., no sharp edges). Be sure that all recycled materials are clean.

You may wish to enlist the assistance of adult volunteers or older students to work with teams as they move through the Plan, Try, and Test steps of the EDP in the Incredible Instruments activity.

Students will present their musical performances as a culminating activity. If you wish to invite outside guests such as parents, caregivers, or other students, make appropriate preparations. Plan to have a clear space in the classroom for students to display their team name and present their musical performances. Be prepared with appropriate equipment if you wish to video record students' performances.

LEARNING COMPONENTS
Introductory Activity/Engagement

Connection to the Challenge: Begin each day of this lesson by directing students' attention to the module challenge, the Show Me the Waves Challenge:

> *You and your team will be challenged to create a musical show that demonstrates how waves can be used to communicate over distances. You will create musical instruments that create sound waves, and you will also use light to show how sound and light can be used to communicate and entertain.*

Hold a brief class discussion of how students' learning in the previous days' lessons contributed to their ability to complete the challenge and how students are progressing through their work on the challenge. Remind students each day of the collaboration rules the class compiled in Lesson 2.

Science Class: Hold a class discussion about using sound and light in musical performances, asking students the following:

- How is sound used in musical performances?

- How are sound waves created in musical performances?

- How can light be used in a performance?

- Have you ever attended a musical performance? If so, how were sound and light used?

- How do we usually experience music (e.g., in person or through technology that allows it to be communicated over long distances)?

Introduce to students the song or songs you chose for their performance or, alternatively, if you chose to have students write their own songs, provide guidelines for this (e.g., the song can be no longer than 1 minute).

Mathematics Connection: Introduce to students the idea that mathematical concepts are necessary for planning a musical performance. Tell students that they will have a total of 3 minutes for their musical performance and that during that time they will need to set up for their performance, introduce their team, perform their song, and present their light elements. Emphasize that their song should last for 1 minute, but that their whole performance should not last longer than 3 minutes. Hold a discussion about how to plan for time in a musical performance, asking students for their ideas about how much time might be required for each task. Next, tell students that they are going to guess how long 1 minute is without looking at a clock. Instruct students to close their eyes and, once you give the start signal, tell students that you are going to set your timer for 1 minute. Ask students to estimate when they think 1 minute has passed and raise their hands at that point. Record the number of students who raise their hand before 1 minute has passed. Ask students if they were surprised at how long or how short a minute seemed. Next, choose a song to sing as a class and set your timer for 1 minute. As a class, sing the song for 1 minute (you may have to sing the song more than once).

ELA Connection: Hold a class discussion about students' experiences with musical instruments, asking students what instruments they know about and what instruments they have heard. Conduct an interactive read-aloud of *I Know a Shy Fellow Who Swallowed a Cello,* by Barbara S. Garriel, asking students to be alert to the musical instruments in the story. After the read-aloud, hold a class discussion about what students learned about musical instruments. Remind students that in Lesson 2 they made musical instruments that vibrate when struck, like a drum. Tell students that drums are a type of instrument called percussion instruments. Tell students that string instruments are another type of instrument. Ask students to name examples of string instruments (e.g., cello, guitar,

violin) and tell students they are going to make string instruments in this lesson to use in their musical performance.

Social Studies Connection: Remind students that in Lesson 2 they learned about how people who have hearing impairments can experience music through vibrations and that in Lesson 3 they learned about how people who have visual impairments can experience visual images through others' descriptions. Tell students that in this lesson they are going to learn about a person who had both hearing loss and vision loss and who learned to experience the world in unique ways and even wrote books. Conduct an interactive read-aloud of *A Picture Book of Helen Keller,* by David A. Adler.

Activity/Exploration

Science Class and Mathematics Connection: Students will work in the teams established in Lesson 2 to create additional musical instruments for their performance. Students will create stringed instruments (guitars) in the Singing Strings activity and will use the EDP to create their own innovative instruments in the Incredible Instruments activity. Student teams will use these instruments and the drums they created in Lesson 2 (I Can Hear the Music activity) to stage their musical performances.

Singing Strings Activity

Students will use the POE process for the Singing Strings activity. Create a POE chart. Students will also create STEM Research Notebook entries for each step of the process. Group students into their teams for this activity and have students work as pairs within their teams. If there are three students on a team, have students work to create two tissue box guitars.

As you prepare for the Singing Strings activity, show students the materials they will use to create their instruments (tissue boxes, paper towel rolls, various widths of rubber bands). Next, show students the completed model of a tissue box guitar you prepared and demonstrate how the instrument makes sound. Point out to students that all the rubber bands on the guitar you created are the same size but that they will be able to use rubber bands of different widths and lengths. Ask students to make predictions about how the widths of rubber bands will affect the sound of the instruments by asking the following:

- Will the width of the rubber band affect the sound? (i.e., will it be higher pitch, lower pitch?)

- Will the length of the rubber band affect the sound? (i.e., will it be higher pitch, lower pitch?)

Record students' responses on a KLEWS chart, and have students record their predictions in STEM Research Notebook Entry #30.

STEM Research Notebook Entry #30

Have students make predictions about how the lengths and widths of rubber bands will affect the sound of the guitars they will create.

Next, hand out all the activity materials to each student team. There are materials provided for teams to split into pairs to each create a tissue box guitar. For teams of three, students can work together to make two guitars.

Instruct students to hold the paper towel roll against the short side of the tissue box and trace the outline of the roll. Next, students should cut a hole where they have traced and insert the paper towel roll to serve as a handle. Have students glue or tape the paper towel roll in place. Next, tell students to look at their four rubber bands' lengths and thicknesses. Students should identify long and short rubber bands and should also note whether the rubber band is thick, thin, or medium width. They should then stretch the rubber bands around the long side of the box, leaving a small space between the rubber bands. Have students test each of the rubber band strings and record their observations in STEM Research Notebook Entry #31.

STEM Research Notebook Entry #31

Have students document their observations about the effect of lengths and widths of rubber bands on the sound of their guitars in their STEM Research Notebooks.

Hold a class discussion, asking whether the width of the rubber bands affected the sound. Add student responses to the POE chart.

Revisit student predictions and discuss whether they were accurate or not. Add student responses to the POE chart. As a class, have students offer their ideas about why different rubber bands sound different, and guide students to understand that thinner and shorter rubber bands vibrate more quickly than wider and longer rubber bands and that we hear the faster vibrations as a higher pitched sound. Add students' responses to the POE chart, and have students record their explanations in STEM Research Notebook Entry #32.

STEM Research Notebook Entry #32

Have students document their cause-and-effect explanations in their STEM Research Notebooks.

Incredible Instruments

Introduce the Incredible Instruments activity by telling students that now they have drums (from Lesson 2) and guitars (from the Singing Strings activity) for their musical performance. Now, students will have the chance to use their creativity to design and make another instrument of their choosing for their performance. Remind students of the steps of the EDP.

First, as a class, work through the Define and Learn steps of the EDP, and record students' responses on chart paper. For the Learn step of the EDP, prompt students to understand that they need to know what materials are available and how much time they have to create their instruments.

Next, for the Plan step, give each team of students a set of recycled materials as well as masking tape, glue, and craft materials. Tell students that they should plan and sketch their instruments before they start to build them. Have students each create labeled sketches of their idea for their team's instrument in STEM Research Notebook Entry #33. Encourage students to use their creativity and what they know about sound waves to create their instruments.

STEM Research Notebook Entry #33

Have students create labeled sketches of their ideas for their team's instrument in their STEM Research Notebooks.

After each student has completed a sketch of an instrument, have each student present his or her sketch to the team and have teams decide on one instrument to build. As teams build their instruments, remind them to use the sketch. After teams have built their instruments, have them test the instruments to see if they make sounds as intended. Hold a class discussion about whether teams' instruments turned out as intended and what, if anything, they can do to improve them.

ELA Connection: Have each student team work together to devise a name for their team to use during their musical performance. Ask students what the purpose is for having a name for a musical group, and create a list of students' responses (e.g., easier to have one name than to list each person's name, tells something about the kind of music the group plays, tells something about the people in the group). Remind students that they learned about adjectives and imagery in this module and how words can provide important information. Provide students with some examples of current popular music groups with which they may be familiar. Ask students to count how many words each of these names use. Emphasize to students that group names usually use only a few words so that they are easy to remember, but that these words are descriptive and memorable.

Challenge groups to identify names for their groups that tell something about their team while using only a few words.

Social Studies Connection: Hold a class discussion about how students think they will react when they encounter people who are differently abled (e.g., those with blindness or deafness). Introduce the term *empathy* and hold a class discussion about what it means to try to put yourself in someone else's situation. Conduct an interactive read-aloud of *Stand in My Shoes: Kids Learning About Empathy,* by Bob Sornson. Hold a class discussion about what they learned about empathy in the story. Ask students for their ideas about how they could be more empathetic with people in their lives. Next, ask students how they could use this information with people who have visual impairments or hearing loss.

Explanation

Science Class and Mathematics Connection: Have student teams choose one instrument for each student to use during the performance. Next, show the students the lighting materials available (see materials list on p. 107). Pass out the lighting materials to teams and allow teams to experiment with them. Hold a class discussion about whether the materials are transparent, translucent, or opaque. Next, have teams decide how to incorporate the light elements into their performances (e.g., a flashlight on a desk pointed toward them or students who can operate a musical instrument with one hand can hold a flashlight with the other).

Once teams have decided on their instruments and how to incorporate light, have teams practice their performances several times using stopwatches to ensure that they stay within their time limit. Remind teams of their time limit of 3 minutes to set up, introduce themselves, and present their performances.

ELA Connection: Have each team create a sign on poster board with the team name.

Social Studies Connection: Hold a class discussion about appropriate audience behavior at a musical performance, connecting the idea of audience behavior and empathy with performers.

Elaboration/Application of Knowledge

Science Class and ELA, Mathematics, and Social Studies Connections: Hold a class performance day, giving each team 3 minutes to set up, introduce their team, and present their musical performances. Before the performances begin, remind students about empathy and how they can use empathy as an audience. After each team has performed, hold a class discussion about the different sounds, how they were made, and how the audience heard them. Also hold a class discussion about how the light elements added to the entertainment value of the performances.

Hold a class discussion about how students felt as they were performing. Ask students to share their ideas about whether or not the audience's reactions were helpful to them.

Have students complete the Lesson 4 Assessment in Appendix B (see p. 168).

Evaluation/Assessment

Students may be assessed on the following performance tasks and other measures listed.

Performance Tasks

- Singing Strings

- Incredible Instruments

- Musical performance

- Lesson 4 Assessment (see Appendix B, p. 168)

Other Measures (see assessment rubric in Appendix B, p. 170)

- Teacher observations

- STEM Research Notebook entries

- Participation in teams during investigations

INTERNET RESOURCES

Not applicable.

SUGGESTED BOOKS

- *Duke Ellington: The Piano Prince and His Orchestra* by Andrea Pinkney (Hyperion Books for Children, 1998)

- *JAZZ on a Saturday Night* by Leo Dillon and Diane Dillon.(Blue Sky Press, 2007)

- *Light: Science Secrets* by Jason Cooper (Rourke Corporation, 1992)

- *Listen to the City* by Rachel Isadora (Putnam, 2000)

- *Max Found Two Sticks* by Brian Pinkney (Simon & Schuster, 1997)

- *Oscar and the Bat: A Book About Sound* by Geoff Waring (Candlewick Press, 2008)

- *Oscar and the Moth: A Book About Light and Dark* by Geoff Waring (Candlewick Press, 2006)

- *Punk Farm* by Jarrett Krosoczka (Alfred Knopf, 2005).

- *Punk Farm on Tour* by J. Krosoczka (Alfred Knopf, 2007)

- *Sound and Light* by Mike Goldsmith (Kingfisher, 2007)

- *Sound: Loud, Soft, High, and Low* by Natalie Rosinksy (Picture Window Books, 2002)

- *STEM Jobs in Movies* by Carla Mooney (Rourke Educational Media, 2014)

- *STEM Jobs in Music* by Shirley Duke (Rourke Educational Media, 2014)

- *The Magic School Bus in the Haunted Museum: A Book About Sound* by Linda Beech (Scholastic, 1995)

- *What Are Light Waves?* by Robin Johnson (Crabtree Publishing Company, 2014)

- *Zin! Zin! Zin! A Violin* by Lloyd Moss (Simon & Schuster, 1995)

REFERENCE

Koehler, C., M. A. Bloom, and A. R. Milner. 2015. The STEM road map for grades K–2. In *STEM road map: A framework for integrated STEM education*, ed. C. C. Johnson, E. E. Peters-Burton, and T. J. Moore, 41–67. New York: Routledge. *www.routledge.com/products/9781138804234.*

TRANSFORMING LEARNING WITH INFLUENCE OF WAVES AND THE *STEM ROAD MAP CURRICULUM SERIES*

Carla C. Johnson

This chapter serves as a conclusion to the Influence of Waves integrated STEM curriculum module, but it is just the beginning of the transformation of your classroom that is possible through use of the *STEM Road Map Curriculum Series.* In this book, many key resources have been provided to make learning meaningful for your students through integration of science, technology, engineering, and mathematics, as well as social studies and English language arts, into powerful problem- and project-based instruction. First, the Influence of Waves curriculum is grounded in the latest theory of learning for students in grade 1 specifically. Second, as your students work through this module, they engage in using the engineering design process (EDP) and build prototypes like engineers and STEM professionals in the real world. Third, students acquire important knowledge and skills grounded in national academic standards in mathematics, English language arts, science, and 21st century skills that will enable their learning to be deeper, retained longer, and applied throughout, illustrating the critical connections within and across disciplines. Finally, authentic formative assessments, including strategies for differentiation and addressing misconceptions, are embedded within the curriculum activities.

The Influence of Waves curriculum in the Cause and Effect STEM Road Map theme can be used in single-content classrooms (e.g., science) where there is only one teacher or expanded to include multiple teachers and content areas across classrooms. Through the exploration of the Show Me the Waves Challenge, students engage in a real-world STEM problem on the first day of instruction and gather necessary knowledge and skills along the way in the context of solving the problem.

The other topics in the *STEM Road Map Curriculum Series* are designed in a similar manner, and NSTA Press has additional volumes in this series for this and other grade levels and plans to publish more. The volumes covering Innovation and Progress have been published and are as follows:

- *Amusement Park of the Future, Grade 6*

- *Construction Materials, Grade 11*

- *Harnessing Solar Energy, Grade 4*

- *Transportation in the Future, Grade 3*

- *Wind Energy, Grade 5*

The volumes covering the Represented World have also been published and are as follows:

- *Car Crashes, Grade 12*

- *Improving Bridge Design, Grade 8*

- *Investigating Environmental Changes, Grade 2*

- *Packaging Design, Grade 6*

- *Patterns and the Plant World, Grade 1*

- *Radioactivity, Grade 11*

- *Rainwater Analysis, Grade 5*

- *Swing Set Makeover, Grade 3*

The tentative list of other books includes the following themes and subjects:

- Cause and Effect
  - Earth on the move
  - Healthy living
  - Human impacts on our climate
  - Natural hazards
  - Physics in motion
- Sustainable Systems
  - Composting: Reduce, reuse, recycle
  - Creating global bonds

- Hydropower efficiency

- System interactions

- Optimizing the Human Experience

  - Genetically modified organisms

  - Mineral resources

  - Rebuilding the natural environment

  - Water conservation: Think global, act local

If you are interested in professional development opportunities focused on the STEM Road Map specifically or integrated STEM or STEM programs and schools overall, contact the lead editor of this project, Dr. Carla C. Johnson (*carlacjohnson@ncsu.edu*), associate dean and professor of science education and executive director of the William and Ida Friday Institute at North Carolina State University. Someone from the team will be in touch to design a program that will meet your individual, school, or district needs.

APPENDIX A

STEM RESEARCH NOTEBOOK TEMPLATES

MY STEM RESEARCH NOTEBOOK

THE INFLUENCE OF WAVES

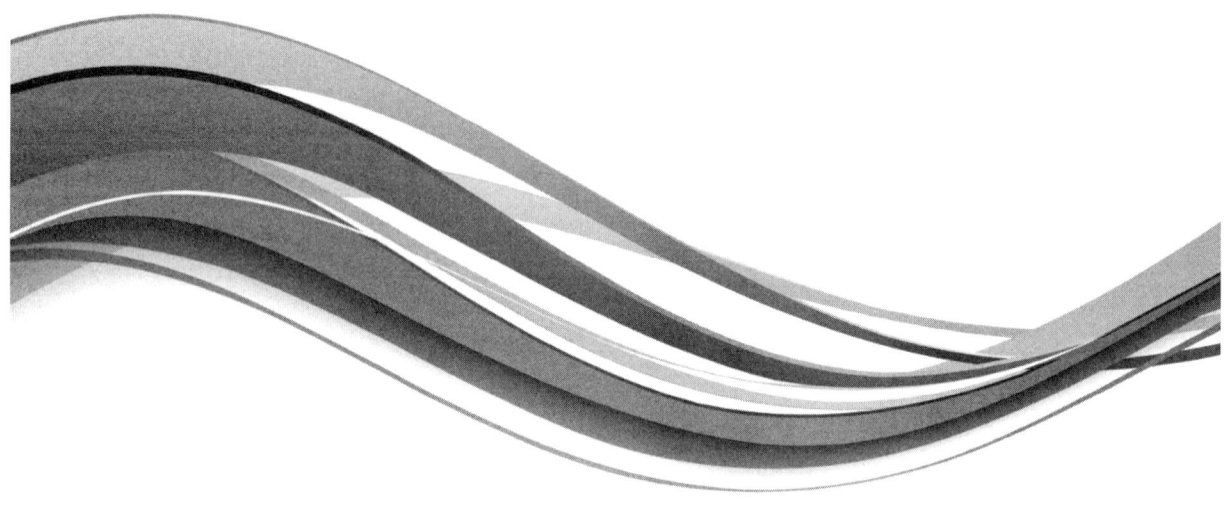

Name:

Name: _____   Date: _____

STEM RESEARCH NOTEBOOK ENTRY #1 (LESSON PLAN 1)

Draw and label what you know about waves.

Name: _____   Date: _____

STEM RESEARCH NOTEBOOK ENTRY #2 (LESSON PLAN 1)

I learned ...

Name: _____

Date: _____

STEM RESEARCH NOTEBOOK ENTRY #3 (LESSON PLAN 1)

VOCABULARY WORDS

Word	Definition	Illustration

Name: _____   Date: _____

STEM RESEARCH NOTEBOOK ENTRIES #4 AND #5 (LESSON PLAN 1)

WAVES IN MOTION

Circle or write in your predictions. Then, circle or write in your observations.

Questions	Predictions	Observations
How can you make waves?	You cannot make waves. Tilting pan Blowing on water Making sound Other: _____ _____	We could not make waves. Tilting pan Blowing on water Making sound Other: _____ _____
How can you make different sizes of waves?	You cannot make different sizes of waves. Tilting pan at different levels Blowing differently Making different sounds Other: _____ _____	We could not make different sizes of waves. Tilting pan at different levels Blowing differently Making different sounds Other: _____ _____
How will bigger waves affect the cork (boat)?	No effect Bigger waves will make the boat: _____ _____	No effect Bigger waves made the boat: _____ _____

Name: _____   Date: _____

STEM RESEARCH NOTEBOOK ENTRY #6 (LESSON PLAN 1)

Measure your wave's height and wavelength:

Our wave amplitude is _____ blocks.

Our wavelength is _____ blocks.

Count the number of crests and troughs on your wave:

There are _____ crests on our wave.

There are _____ troughs on our wave.

Compared with the other team's wave, our wave has:

A _____ (taller or shorter) height.

A _____ (longer or shorter) wavelength.

(More or Fewer) _____ crests.

(More or Fewer) _____ troughs.

Name: _____   Date: _____

STEM RESEARCH NOTEBOOK ENTRY #7 (LESSON PLAN 1)

Adjectives are words that describe a person, a place, or a thing. List adjectives that describe the picture your teacher shows you:

_____

- -

_____

- -

_____

_____

- -

_____

_____

- -

_____

_____

- -

_____

- -

_____

_____

- -

_____

Name: _____   Date: _____

STEM RESEARCH NOTEBOOK ENTRY #8 (LESSON PLAN 1)

EXPLANATIONS

Observation	Cause	Effect
You made waves.		
You made some big waves and some small waves.		
Your cork (boat) moved differently in different waves.		

Name: _____   Date: _____

STEM RESEARCH NOTEBOOK ENTRY #9, PAGE 1 (LESSON PLAN 2)

Answer the following questions with pictures and words:

1. What are things we can hear?

_____
- -
_____

_____
- -
_____

2. How do we hear?

_____
- -
_____

_____
- -
_____

Name: _____   Date: _____

STEM RESEARCH NOTEBOOK ENTRY #9, PAGE 2 (LESSON PLAN 2)

3. How does the sound of a voice move from one person's mouth to another person's ears?

Name: _____   Date: _____

STEM RESEARCH NOTEBOOK ENTRY #10 (LESSON PLAN 2)

Record your observations using pictures and words:

The plastic fork did the following when I made it vibrate:

_____

- -

_____

_____

- -

_____

_____

- -

_____

When I held more of the handle of the fork on the table, the following happened (circle your observations):

The fork vibrated faster.   The fork vibrated slower.   The fork vibrated about the same.

The sound was higher.   The sound was lower.   The sound was about the same.

Name: _____ Date: _____

STEM RESEARCH NOTEBOOK ENTRY #11 (LESSON PLAN 2)

My team used these words to describe the sound:

_____

- -

_____

_____

- -

_____

_____

- -

_____

_____

- -

_____

Draw a picture to describe your team's sound:

Name: _____   Date: _____

STEM RESEARCH NOTEBOOK ENTRY #12 (LESSON PLAN 2)

Use the word bank to label the parts of the ear that help us hear:

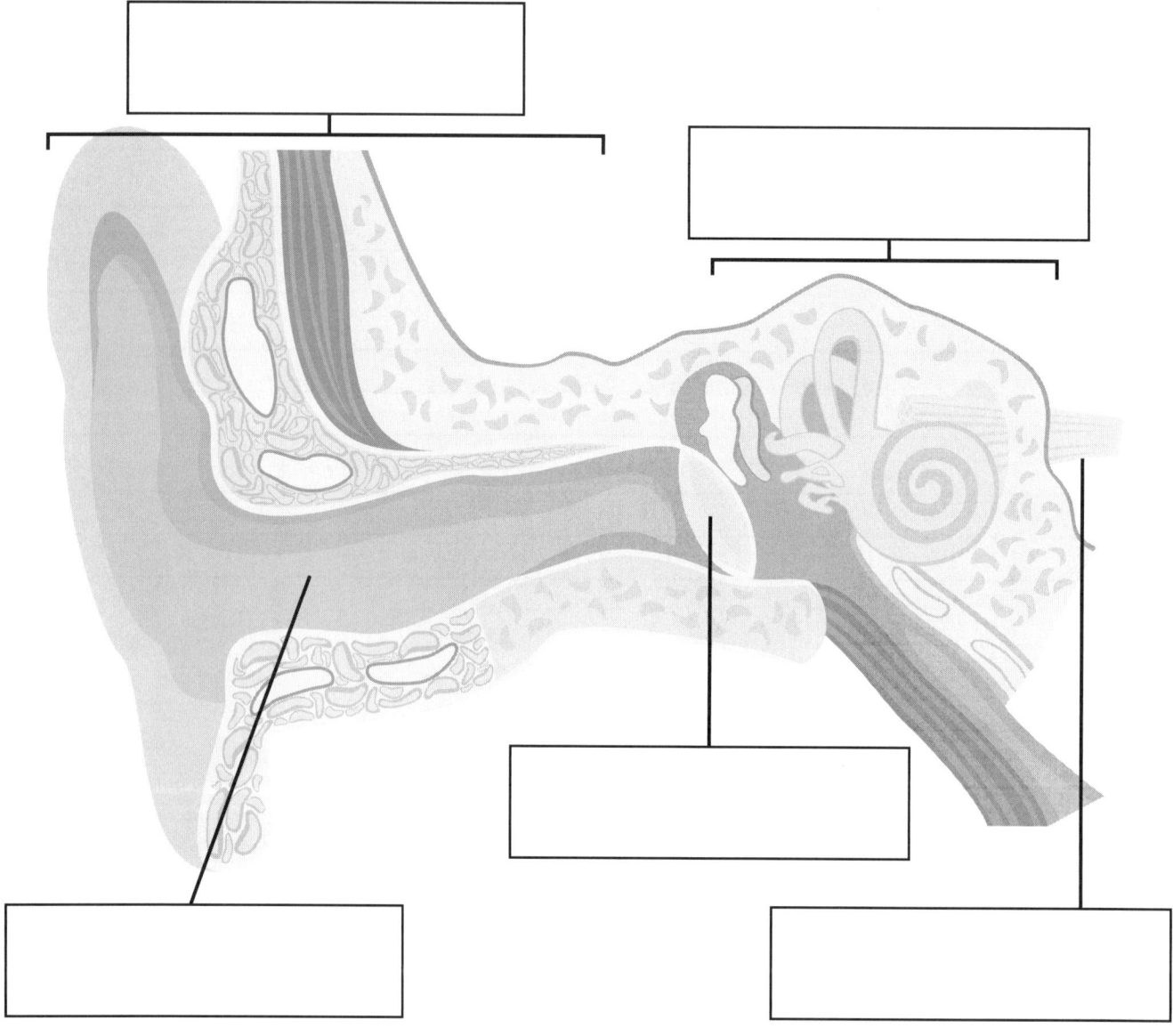

Word Bank

ear canal   outer ear   eardrum

nerves   inner ear

Name: _____   Date: _____

STEM RESEARCH NOTEBOOK ENTRY #13 (LESSON PLAN 2)

I learned ...

_____

- -

_____

- -

_____

- -

_____

- -

_____

Name: _____   Date: _____

STEM RESEARCH NOTEBOOK ENTRIES #14 AND #15, PAGE 1 (LESSON PLAN 2)

WHAT'S THE BUZZ?

Circle your predictions. Then, circle your observations.

Questions	Predictions	Observations
Will the length of the string affect the loudness of the sound?	When you use string longer than 10 feet, your partner's voice will be: Softer The Same Louder	When you used string longer than 10 feet, your partner's voice was: Softer The Same Louder
	When you use string shorter than 10 feet, your partner's voice will be: Softer The Same Louder	When you used string shorter than 10 feet, your partner's voice was: Softer The Same Louder
Will the length of the string affect the pitch of the sound?	When you use string longer than 10 feet, your partner's voice will be: Lower Pitched The Same Higher Pitched	When you used string longer than 10 feet, your partner's voice was: Lower Pitched The Same Higher Pitched
	When you use string shorter than 10 feet, your partner's voice will be: Lower Pitched The Same Higher Pitched	When you used string shorter than 10 feet, your partner's voice was: Lower Pitched The Same Higher Pitched

Name: _____   Date: _____

STEM RESEARCH NOTEBOOK ENTRIES #14 AND #15, PAGE 2 (LESSON PLAN 2)

WHAT'S THE BUZZ?

Circle your predictions. Then, circle your observations.

Questions	Predictions	Observations
Will the tightness of the string affect the loudness of the sound?	When your string is loose compared with when your string is tight, your partners voice will be: Softer The Same Louder	When your string was loose compared with when your string was tight, your partners voice was: Softer The Same Louder
Will the tightness of the string affect the pitch of the sound?	When your string is loose compared with when your string is tight, your partners voice will be: Lower Pitched The Same Higher Pitched	When your string was loose compared with when your string was tight, your partners voice was: Lower Pitched The Same Higher Pitched

Name: _____   Date: _____

STEM RESEARCH NOTEBOOK ENTRY #16 (LESSON PLAN 2)

Cause	Effect	Explanations
When the string was longer ...	the volume of the sound was (softer, the same, louder): _____	I think this is because:
When the string was longer ...	the pitch of the sound was (lower, the same, higher): _____	I think this is because:
When the string was looser ...	the volume of the sound was (softer, the same, louder): _____	I think this is because:
When the string was looser ...	the pitch of the sound was (lower, the same, higher): _____	I think this is because:

Name: _____

Date: _____

STEM RESEARCH NOTEBOOK ENTRY #17 (LESSON PLAN 2)

CLASSROOM SOUNDS

Type of Sound	Estimate of Loudness (in decibels)	Actual Loudness (in decibels)	Difference Between My Estimate and Actual Decibels Smaller: < Same: = Larger: >
	_____ decibels	_____ decibels	
	_____ decibels	_____ decibels	
	_____ decibels	_____ decibels	
	_____ decibels	_____ decibels	
	_____ decibels	_____ decibels	
	_____ decibels	_____ decibels	

Name: _____   Date: _____

STEM RESEARCH NOTEBOOK ENTRY #18, PAGE 1 (LESSON PLAN 2)

Draw and label the first instrument your team decided to make.

Instrument #1

Name: _____   Date: _____

STEM RESEARCH NOTEBOOK ENTRY #18, PAGE 2 (LESSON PLAN 2)

Draw and label the second instrument your team decided to make.

Instrument #2

Name: _____   Date: _____

STEM RESEARCH NOTEBOOK ENTRY #19, PAGE 1 (LESSON PLAN 3)

Answer the following questions with pictures and words:

1. How does light help us see?

_____
- -
_____

_____
- -
_____

2. How do we see?

_____
- -
_____

_____
- -
_____

Name: _____   Date: _____

STEM RESEARCH NOTEBOOK ENTRY #19, PAGE 2 (LESSON PLAN 3)

3. Where does light come from?

_____

_____

_____

_____

Name: _____   Date: _____

STEM RESEARCH NOTEBOOK ENTRY #20 (LESSON PLAN 3)

I learned ...

NATIONAL SCIENCE TEACHING ASSOCIATION

Name: _____   Date: _____

STEM RESEARCH NOTEBOOK ENTRY #21 (LESSON PLAN 3)

The person who invented the lightbulb was

_____

- -

_____

Name: _____   Date: _____

STEM RESEARCH NOTEBOOK ENTRIES #22 AND #23 (LESSON PLAN 3)

LIGHTING THE WAY

Circle your predictions. Then, circle your observations.

Questions	Predictions	Observations
Can you see the picture through clear plastic when you shine a flashlight on it?	Yes No	Yes No
Can you see the picture through wax paper when you shine a flashlight on it?	Yes No	Yes No
Can you see the picture through cardboard when you shine a flashlight on it?	Yes No	Yes No

Name: _____   Date: _____

LIGHTING THE WAY
—OBSERVATIONS—

1. The clear plastic is (circle one):

                Transparent   Translucent   Opaque

2. The wax paper is (circle one):

                Transparent   Translucent   Opaque

3. The cardboard is (circle one):

                Transparent   Translucent   Opaque

Name: _____   Date: _____

STEM RESEARCH NOTEBOOK ENTRIES #25 AND #26 (LESSON PLAN 3)

SHADY SHADOW PUPPETS

Circle your predictions. Then, circle your observations.

Questions	Predictions	Observations
Does the size of the puppet affect the shadow size?	When all the puppets are 10 feet from the wall All the puppets' shadows will be the same size. The biggest puppet's shadow will be larger than the other puppets' shadows. The biggest puppet's shadow will be smaller than the other puppets' shadows.	When all the puppets were 10 feet from the wall All the puppets' shadows were the same size. The biggest puppet's shadow was larger than the other puppets' shadows. The biggest puppet's shadow was smaller than the other puppets' shadows.
Does the distance from the wall affect the shadow size?	When you hold a puppet 10 feet from the wall, its shadow will be Less than: < Equal to: = Greater than: > compared with when you hold the puppet 3 feet from the wall.	When you held a puppet 10 feet from the wall, its shadow was Less than: < Equal to: = Greater than: > compared with when you held the puppet 3 feet from the wall.
Does the shadow get bigger or smaller when the puppet is closer to the wall and further away from the light source?	When you hold your flashlight and puppet 10 feet away from the wall and then move the puppet closer to the wall without moving the flashlight, the puppet's shadow will become Bigger Smaller Stay the same	When you held your flashlight and puppet 10 feet away from the wall and then moved the puppet closer to the wall without moving the flashlight, the puppet's shadow became Bigger Smaller Stayed the same

Name: _____   Date: _____

STEM RESEARCH NOTEBOOK ENTRY #27 (LESSON PLAN 3)

Use the word bank to fill in the blanks.

1. I learned that we see with our _____.

2. We see things when _____ bounces off objects and into our eyes.

3. Light enters our eyes through this part of the eye: _____ .

4. The colored part of my eye is called the _____ .

5. Messages from my eyes go to my _____ so I can know what I am seeing.

Word Bank

pupil   brain   iris   light   eyes

Name: _____   Date: _____

STEM RESEARCH NOTEBOOK ENTRY #28 (LESSON PLAN 3)

I learned ...

_____

- -

_____

_____

- -

_____

I can use what I know about light in a musical show by ...

_____

- -

_____

_____

- -

_____

Name: _____   Date: _____

STEM RESEARCH NOTEBOOK ENTRY #29 (LESSON PLAN 3)

Ask your partner to describe with words the picture he or she has. Draw a picture based on that description.

Name: _____   Date: _____

STEM RESEARCH NOTEBOOK ENTRIES #30 AND #31 (LESSON PLAN 4)

SINGING STRINGS

Circle your predictions. Then, circle your observations.

Questions	Predictions	Observations
How will the length of the rubber band affect the guitar's sound?	A shorter rubber band will sound: Lower Pitched: < The Same: = Higher Pitched: > when compared with a longer rubber band.	A shorter rubber band sounded: Lower Pitched: < The Same: = Higher Pitched: > when compared with a longer rubber band.
How will the width of the rubber band affect the guitar's sound?	A narrower rubber band will sound: Lower Pitched: < The Same: = Higher Pitched: > when compared with a wider rubber band.	A narrower rubber band sounded: Lower Pitched: < The Same: = Higher Pitched: > when compared with a wider rubber band.

Name: _____   Date: _____

STEM RESEARCH NOTEBOOK ENTRY #32 (LESSON PLAN 4)

EXPLANATIONS

Variable	Cause	Effect on Pitch
Length of rubber bands	Shorter rubber bands compared with longer rubber bands vibrate: More slowly The same as More quickly	And this means the pitch we hear is: Higher Lower
Width of rubber bands	Thinner rubber bands compared with thicker rubber bands vibrate: More slowly The same as More quickly	And this means the pitch we hear is: Higher Lower

Name: _____   Date: _____

STEM RESEARCH NOTEBOOK ENTRY #33 (LESSON PLAN 4)

INCREDIBLE INSTRUMENTS

Sketch your idea for your team's musical instrument. Label the materials.

This instrument will make music by

_____

_____

_____

_____

APPENDIX B

LESSON ASSESSMENTS AND ASSESSMENT RUBRIC

Name: _____   Date: _____

ASSESSMENT FOR LESSON PLAN 1

1. Draw and label a picture of waves in water.

2. What is a wave?

_____

_____

_____

_____

3. How did you cause waves in water?

_____

_____

_____

_____

Name: _____   Date: _____

ASSESSMENT FOR LESSON PLAN 2, PAGE 1

1. Draw and label two things that make sound.

Name: _____   Date: _____

ASSESSMENT FOR LESSON PLAN 2, PAGE 2

2. How do sound waves travel to our ears?

_____

- -

_____

_____

- -

_____

3. Label two parts of the human ear in the picture:

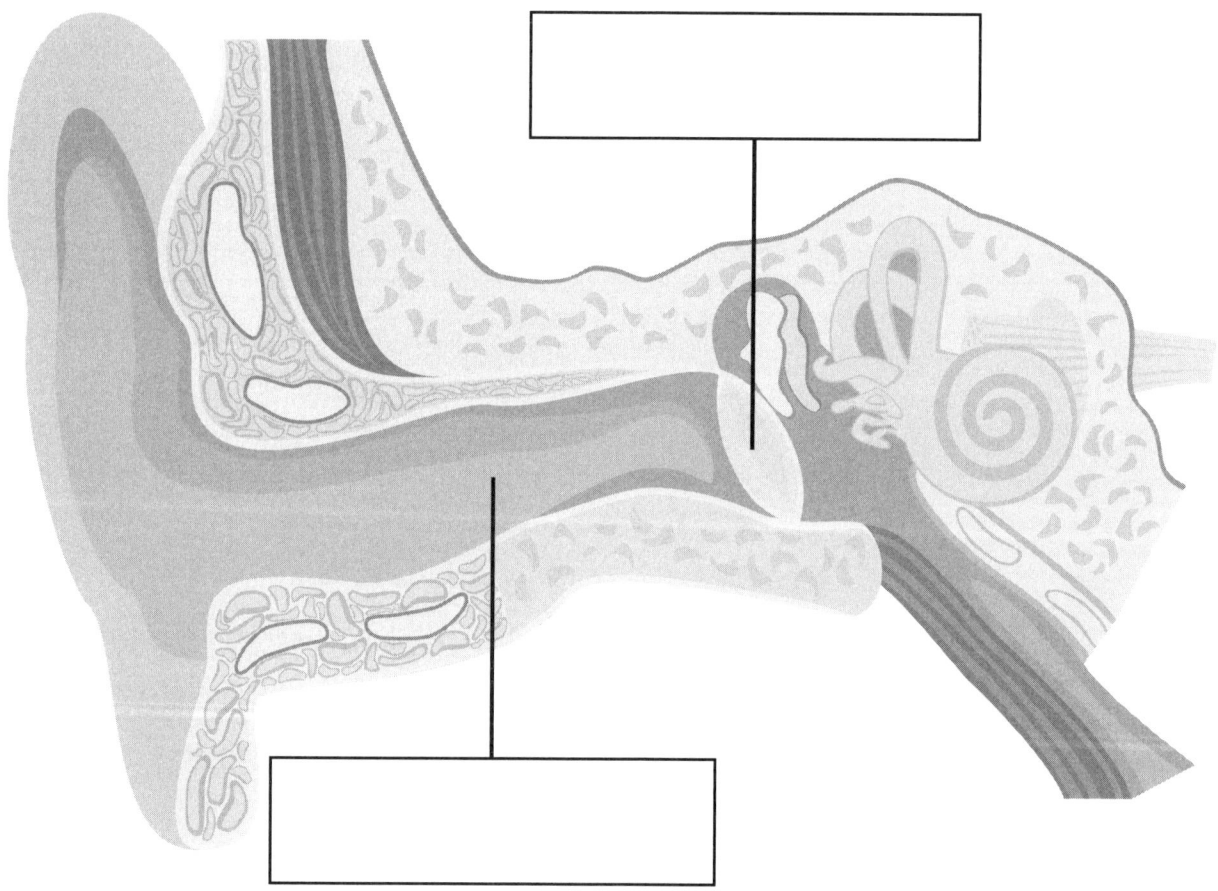

Name: _____   Date: _____

ASSESSMENT FOR LESSON PLAN 3, PAGE 1

Draw and label two things we can see because of light, and explain why we can see these things.

1.

_____

_____

_____

_____

Name: _____   Date: _____

ASSESSMENT FOR LESSON PLAN 3, PAGE 2

2.

Name: _____   Date: _____

ASSESSMENT FOR LESSON PLAN 3, PAGE 3

3. Draw and label one thing that is transparent.

4. Draw and label one thing that is translucent.

5. Draw and label one thing that is opaque.

Name: _____   Date: _____

ASSESSMENT FOR LESSON PLAN 4, PAGE 1

1. Using pictures and words, explain how you created sound waves in your team's musical performance.

2. Using pictures and words, explain how your audience heard your music in your team's musical performance.

Name: _____   Date: _____

ASSESSMENT FOR LESSON PLAN 4, PAGE 2

3. Draw and name one other way your team used sound to communicate during this module.

_____

- -

_____

_____

- -

_____

4. Using pictures and words, describe how you used light in your team's musical performance and what kind of materials you used.

_____

- -

_____

_____

- -

_____

5. Were the materials you used to create light transparent, translucent, or opaque?

_____

- -

_____

Observation, STEM Research Notebook, and Participation Rubric

Name: _____

Categories (components)	Missing or Unrelated (0 points)	Beginning (1 point)	Developing (2 points)	Meets Expectations (3 points)	Exceeds Expectations (4 points)	Score
OBSERVATION OF LISTENING AND DISCUSSION SKILLS	Component is missing or unrelated.	Does not listen to others and shows little respect for alternative viewpoints.	Occasionally listens to others but often speaks out of turn.	Listens to others, only occasionally speaks out of turn, and generally accepts other points of view.	Listens carefully to others, waits for turn to speak, and respects alternative viewpoints.	
STEM RESEARCH NOTEBOOK	Component is missing or unrelated.	Indicates little understanding of the concepts being taught.	Recalls and is able to explain basic facts and concepts.	Demonstrates ability to apply concepts, using information in new situations.	Demonstrates a deep understanding of concepts by drawing relationships between ideas and using information to generate new ideas.	
PARTICIPATION	Component is missing.	Does not volunteer. When responding to teacher prompts, comments are sometimes not relevant to the discussion.	Responds to teacher prompts during classroom discussions but does not volunteer.	Willingly participates in classroom discussions and offers relevant comments.	Contributes insightful comments and poses thoughtful questions.	

TOTAL SCORE: _____

COMMENTS:

APPENDIX C

CONTENT STANDARDS ADDRESSED IN THIS MODULE

NEXT GENERATION SCIENCE STANDARDS

Table C1 (p. 172) lists the science and engineering practices, disciplinary core ideas, and crosscutting concepts this module addresses. The supported performance expectations are as follows:

- 1-PS4-1. Plan and conduct investigations to provide evidence that vibrating materials can make sound and that sound can make materials vibrate.

- 1-PS4-2. Make observations to construct an evidence-based account that objects can be seen only when illuminated.

- 1-PS4-3. Plan and conduct an investigation to determine the effect of placing objects made with different materials in the path of a beam of light.

- 1-PS4-4. Use tools and materials to design and build a device that uses light or sound to solve the problem of communicating over a distance.

Table C1. *Next Generation Science Standards (NGSS)*

Science and Engineering Practices

PLANNING AND CARRYING OUT INVESTIGATIONS

Planning and carrying out investigations to answer questions or test solutions to problems in K–2 builds on prior experiences and progresses to simple investigations, based on fair tests, which provide data to support explanations or design solutions.

- Plan and conduct investigations collaboratively to produce evidence to answer a question.

CONSTRUCTING EXPLANATIONS AND DESIGNING SOLUTIONS

Constructing explanations and designing solutions in K–2 builds on prior experiences and progresses to the use of evidence and ideas in constructing evidence-based accounts of natural phenomena and designing solutions.

- Make observations (firsthand or from media) to construct an evidence-based account for natural phenomena.

Disciplinary Core Ideas

PS4.A: WAVE PROPERTIES

- Sound can make matter vibrate, and vibrating matter can make sound.

PS4.B: ELECTROMAGNETIC RADIATION

- Objects can be seen if light is available to illuminate them or if they give off their own light.

- Some materials allow light to pass through them, others allow only some light through and others block all the light and create a dark shadow on any surface beyond them, where the light cannot reach. Mirrors can be used to redirect a light beam. (Boundary: The idea that light travels from place to place is developed through experiences with light sources, mirrors, and shadows, but no attempt is made to discuss the speed of light.)

PS4.C: INFORMATION TECHNOLOGIES AND INSTRUMENTATION

- People also use a variety of devices to communicate (send and receive information) over long distances.

Crosscutting Concept

CAUSE AND EFFECT

- Simple tests can be designed to gather evidence to support or refute student ideas about causes.

Source: NGSS Lead States. 2013. *Next Generation Science Standards: For states, by states.* Washington, DC: National Academies Press. *www.nextgenscience.org.*

Table C2. Common Core Mathematics and English Language Arts (ELA) Standards

MATHEMATICAL PRACTICES	READING STANDARDS
• MP1. Make sense of problems and persevere in solving them. • MP2. Reason abstractly and quantitatively. • MP3. Construct viable arguments and critique the reasoning of others. • MP4. Model with mathematics. • MP5. Use appropriate tools strategically. • MP6. Attend to precision. • MP7. Look for and make use of structure. • MP8. Look for and express regularity in repeated reasoning. **MATHEMATICAL CONTENT** • 1.NBT.A.1. Count to 120, starting at any number less than 120. In this range, read and write numerals and represent a number of objects with a written numeral. • 1.NBT.B.3. Compare two two-digit numbers based on meanings of the tens and ones digits, recording the results of comparisons with the symbols >, =, and <. • 1.MD.C.4. Organize, represent, and interpret data with up to three categories; ask and answer questions about the total number of data points, how many in each category, and how many more or less are in one category than in another. • 1.OA.A.2. Solve word problems that call for addition of three whole numbers whose sum is less than or equal to 20, e.g., by using objects, drawings, and equations with a symbol for the unknown number to represent the problem.	• RI.1.1. Ask and answer questions about key details in a text. • RI.1.2. Identify the main topic and retell key details of a text. • RI.1.3. Describe the connection between two individuals, events, ideas, or pieces of information in a text. • RI.1.7. Use the illustrations and details in a text to describe its key ideas. **WRITING STANDARDS** • W.1.2. Write informative/explanatory texts in which they name a topic, supply some facts about the topic, and provide some sense of closure. • W.1.6. With guidance and support from adults, use a variety of digital tools to produce and publish writing, including in collaboration with peers. • W.1.7. Participate in shared research and writing. • W.1.8. With guidance and support from adults, recall information from experiences or gather information from provided sources to answer a question. **SPEAKING AND LISTENING STANDARDS** • SL.1.1. Participate in collaborative conversations with diverse partners about *grade 1 topics and texts* with peers and adults in small and larger groups. • SL.1.1.A. Follow agreed-upon rules for discussions.

Continued

Table C2. (*continued*)

	SPEAKING AND LISTENING STANDARDS (*continued*)
	• SL.1.1.B. Build on others' talk in conversations by responding to the comments of others through multiple exchanges.
	• SL.1.1.C. Ask questions to clear up any confusion about the topics and texts under discussion.
	• SL.1.3. Ask and answer questions about what a speaker says in order to gather additional information or clarify something that is not understood.
	• SL.1.5. Add drawings or other visual displays to descriptions when appropriate to clarify ideas, thoughts, and feelings.

Source: National Governors Association Center for Best Practices and Council of Chief State School Officers (NGAC and CCSSO). 2010. *Common core state standards.* Washington, DC: NGAC and CCSSO.

Table C3. National Association for the Education of Young Children (NAEYC) Standards

NAEYC Curriculum Content Area for Cognitive Development: Science and Technology
• 2.E.1. Arrange firsthand, meaningful experiences that are intellectually and creatively stimulating, invite exploration and investigation, and engage children's active, sustained involvement by providing a rich variety of material, challenges, and ideas.
• 2.F.3. Extend the range of children's interests and the scope of their thought, present novel experiences and introduce stimulating ideas, problems, experiences, or hypotheses.
• 2.F.6. Enhance children's conceptual understanding through various strategies, including intensive interview and conversation, encourage children to reflect on and "revisit" their experiences.
• 2.G.2. Scaffolding takes on a variety of forms.
• 2.J.1. Incorporate a wide variety of experiences, materials and equipment, and teaching strategies to accommodate the range of children's individual differences in development, skills and abilities, prior experiences, needs, and interests.
• 3.A.1. Teachers consider what children should know, understand, and be able to do across the domains.

Source: National Association for the Education of Young Children (NAEYC). 2005. *NAEYC early childhood program standards and accreditation criteria: The mark of quality in early childhood education.* Washington, DC: NAEYC.

Table C4. 21st Century Skills from the Framework for 21st Century Learning

21st Century Skills	Learning Skills and Technology Tools	Teaching Strategies	Evidence of Success
INTERDISCIPLINARY THEMES	• Economic, Business, and Entrepreneurial Literacy • Health Literacy • Environmental Literacy	• Provide students with the opportunity to investigate waves as they relate to sound in the context of the economics of everyday life (e.g., communicating with sound via technology, natural versus artificial light).	• Students communicate their prior experiences with sound and light in the context of everyday life.
LEARNING AND INNOVATION SKILLS	• Creativity and Innovation • Critical Thinking and Problem Solving • Communication and Collaboration	• Facilitate creativity and innovation through the use of the engineering design process to create and test models.	• Students demonstrate creativity and innovation, critical thinking and problem solving, and communication and collaboration as they develop models to demonstrate how humans experience and interact with sound and light waves and use sound and light to communicate over distances.
INFORMATION, MEDIA, AND TECHNOLOGY SKILLS	• Information Literacy • Media Literacy • Information, Communications, and Technology Literacy	• Engage students in guided practice and scaffolding strategies through the use of developmentally appropriate books, videos, and websites to advance their knowledge.	• Students acquire and use deeper content knowledge as they develop models to demonstrate how humans experience and interact with sound and light waves.
LIFE AND CAREER SKILLS	• Flexibility and Adaptability • Initiative and Self-Direction • Social and Cross-Cultural Skills • Productivity and Accountability • Leadership and Responsibility	• Facilitate student collaborative teamwork to foster life and career skills.	• Throughout this module, students collaborate to develop models to demonstrate how humans experience and interact with sound and light waves.

Source: Partnership for 21st Century Learning, Battelle for Kids. 2015. Framework for 21st Century Learning. *www.battelleforkids.org/networks/p21/frameworks-resources.*

Table C5. English Language Development (ELD) Standards

ELD STANDARD 1: SOCIAL AND INSTRUCTIONAL LANGUAGE

English language learners communicate for Social and Instructional purposes within the school setting.

ELD STANDARD 2: THE LANGUAGE OF LANGUAGE ARTS

English language learners communicate information, ideas, and concepts necessary for academic success in the content area of Language Arts.

ELD STANDARD 3: THE LANGUAGE OF MATHEMATICS

English language learners communicate information, ideas, and concepts necessary for academic success in the content area of Mathematics.

ELD STANDARD 4: THE LANGUAGE OF SCIENCE.

English language learners communicate information, ideas, and concepts necessary for academic success in the content area of Science.

ELD STANDARD 5: THE LANGUAGE OF SOCIAL STUDIES

English language learners communicate information, ideas, and concepts necessary for academic success in the content area of Social Studies.

Source: WIDA. 2012. 2012 amplification of the English language development standards: Kindergarten–grade 12. *https://wida.wisc.edu/teach/standards/eld.*

INDEX

Page numbers printed in **boldface type** indicate tables, figures, or handouts.